Mathematics for economists

Mathematics for economists

A first course

J. M. Pearson

Longman
London and New York

Longman Group Limited
Longman House
Burnt Mill, Harlow, Essex, UK

© Longman Group Limited 1982

*Published in the United States of America
by Longman Inc., New York*

All rights reserved. No part of this publication may be
reproduced, stored in a retrieval system, or transmitted
in any form or by any means, electronic, mechanical,
photocopying, recording, or otherwise, without the
prior permission of the Copyright owner.

First published 1982

British Library Cataloguing in Publication Data

Pearson, J.M.
  Mathematics for economists.
  1. Economics, Mathematical
  I. Title
  510'.2433   HB135

  ISBN 0-582-29615-3

Library of Congress Cataloguing in Publication Data

Pearson, J.M., 1946-
  Mathematics for economists.

  Bibliography: p.
  Includes index.
  1. Economics, Mathematical. I. Title.
  HB135. P42   510'.24339   81-14309
  ISBN 0-582-29615-3   AACR2

Printed in Singapore by Selector Printing Co Pte Ltd

Contents

Preface vii

Chapter 1 Sets 1
1.1 Introduction 1
1.2 The real number system 2
1.3 More about sets 4
1.4 Operations on sets 6
1.5 Venn diagrams 11
1.6 Complements 16
1.7 Applications to economics 18

Chapter 2 Mappings and functions 20
2.1 Introduction 20
2.2 The image set 22
2.3 Types of mappings 23
2.4 Notation 26
2.5 Graphs 28
2.6 More on graphs 33
2.7 Named functions 35
2.8 Reverse mappings and inverse functions 41
2.9 Some more named functions 43
2.10 Applications to economics 53

Chapter 3 Operations on mappings 65
3.1 Introduction 65
3.2 Operations 65
3.3 Composition of functions (function of a function) 68
3.4 Derivation of formula for *gof* 70
3.5 Derivation of formula for *fog* 72

Chapter 4 Equations — 75
4.1 Introduction — 75
4.2 Linear equations — 76
4.3 Quadratic equations — 80
4.4 Types of solution for quadratic equations — 86
4.5 Complex numbers — 90
4.6 Higher degree polynomial equations — 92
4.7 Inequalities — 93
4.8 Simultaneous equations — 97
4.9 Applications to economics — 107

Chapter 5 Differentiation of functions of one variable — 115
5.1 Introduction — 115
5.2 The slope of a function at a point — 117
5.3 A method for finding the slope of a function — 118
5.4 Derivatives and differentiation — 121
5.5 Rules of differentiation — 123
5.6 Applications to economics — 132

Chapter 6 Maximisation and minimisation of functions — 138
6.1 Introduction — 138
6.2 First order condition for a maximum or minimum point — 140
6.3 Stationary points — 146
6.4 Second and higher derivatives — 148
6.5 Second order conditions for maximum and minimum points — 149
6.6 Applications to economics — 154

Chapter 7 Integration — 162
7.1 Introduction — 162
7.2 Rules of integration — 163
7.3 The arbitrary constant of integration — 167
7.4 More difficult rules of integration — 170
7.5 Definite integrals — 177
7.6 Integration and areas under curves — 180
7.7 Areas below the x axis — 182
7.8 Applications to economics — 184

Answers to exercises — 189

Bibliography — 206

Index — 207

Preface

This book is an introductory text in mathematics, designed primarily for undergraduate students of economics, although social science students, required to undertake at least some mathematical training as part of their degree, should also find it useful.

Economists use mathematics extensively to assist them in analysing economic problems, and although some economists lament the increasing use of some of the more high-powered mathematical techniques, which they claim obscure the real economics, it is a fact that mathematical techniques of the type introduced in this book provide an invaluable aid to the understanding and analysis of many economic problems. Consider, for example, a firm faced with deciding how much should be produced to maximise its profit. This is a key problem in economics, which all economists must understand. In order to analyse and solve problems posed by maximisation, the techniques of differential calculus are used. Thus it is only by having a good grasp of mathematics that an economist can successfully master economics. This fact is reflected in the increasing introduction of mathematical and quantitative methods into economic and social science courses.

This book provides a one-term course in mathematics for economists, for students who have progressed only as far as O-level mathematics. Since, in many cases, this was several years ago, the book revises topics covered in an O-level syllabus (sets, graphs, equations etc.) and then introduces other topics students need to grasp in order to cope with an economics degree (differentiation, optimisation, integration). The book is particularly suitable for students possessing only O-level mathematics, who are taking a first-level undergraduate 'quantitative methods' course, consisting of half mathematics and half statistics. These students should find that most of the mathematics in their course is covered in this book. The book will also prove useful to those students, with only O-level mathematics, who are taking a full year's first level mathematics course, but find difficulty, at the beginning of the course, in coping with the texts recommended. This book will provide a 'bridge' between O-level (much of which needs revising) and these more advanced texts.

The book adopts the 'set' or 'modern mathematics' approach, although it covers the same material as in more traditional or 'old mathematics' textbooks. There are two reasons for this approach: firstly because set theory is increasingly being used as a technique in analysing economic problems, and secondly because it provides a unifying approach to the topics covered.

In addition to presenting mathematical techniques, the book attempts to demonstrate some of the applications of these techniques to elementary economic problems, by including at the end of some of the chapters, a section on 'Applications to economics'. These sections illustrate the relevance to economics of the mathematical techniques introduced in the chapter. For example in Chapter 2, where functions are introduced, the section on applications at the end of that chapter discusses demand and supply functions, cost, revenue and profit functions. In Chapter 6, where maximisation and minimisation of functions is covered, the 'Applications to economics' section discusses profit maximisation and the relation between marginal revenue and marginal cost, illustrating how a grasp of the mathematical techniques is invaluable in understanding the economics.

Chapter 1, on sets, does not include a section on 'Applications to economics', not because sets have no applications or uses in economics, but because the groundwork necessary to present useful applications requires more space than could be allocated in this book. (The reader interested in the applicability of sets to economic problems is referred to V. C. Walsh *Introduction to Contemporary Microeconomics*.) Although Chapter 1 would have been included in its own right, as an introduction to the ideas presented in, for example, Walsh, the main reason for its inclusion is that it provides both a foundation and a framework on which the other mathematical techniques can be discussed. As A. C. Chiang states '. . . . the concept of sets underlies every branch of modern mathematics', (A. C. Chiang, *Fundamental Methods of Mathematical Economics*), and it is in this belief that I have included the chapter and used the ideas developed in the chapter as the basis for the techniques discussed in subsequent chapters.

The book has several features which should make it attractive to many students.

Firstly, in keeping with recent development in mathematics teaching in the schools, it begins by introducing sets, a topic most people who have been taught mathematics in school in the last decade are familiar with. Many of the older mathematics for economists books do not mention the topic, and indeed many lecturers in the subject do not cover it, which seems strange to a generation of students nurtured on 'new Maths'.

Secondly, it introduces the mathematical topics as pure mathematics. Many students beginning an introduction to mathematics for economists course will not yet have learned any economics, and will not, initially, be able to apply the mathematical techniques they have learned. These students will be able to use the book leaving out the 'Applications to economics' sections. When they have learned some economics they will be able to go over the ground again, making use of the applications section. On the other hand, many students are reluctant to apply themselves to mastering the mathematical techniques, unless they can see the relevance of doing so. The applications sections are particularly useful for these students, as they can see immediately the relevance of the mathematical topics to economics. The application sections will also be particularly useful to those students who start the course already having some knowledge of economics.

Finally, it is hoped that this book will be especially useful to those students, the majority in many universities and polytechnics, who find it difficult to master mathematical techniques. The book covers every topic thoroughly, explains each point fully and clearly, and moves forward at a pace most students find comfortable. It illustrates points with graphs and examples wherever necessary and provides plenty of exercises for students to try out the techniques for themselves, and consolidate their grasp of the subject by working through problems.

My grateful thanks are extended to George Zis and Professor Michael Sumner, of Salford University and to Jon Stewart of Manchester University, for their encouragement, without which the book would not have been written, and their comments, without which it would have been an inferior publication. Thanks are also due to Mrs M. Ward, Mrs J. M. Robertson and Mrs B. Masters for their typing skills, and finally to my wife, not only for her help in typing and proofreading, but also for the encouragement and comfort she offered throughout the writing of the book.

J. M. Pearson

Chapter 1

Sets

In this chapter we lay the foundations for most of the mathematics which follows in subsequent chapters. We shall not provide economic examples of the use of set theory because of limitations of space. However, students will appreciate the value of the material in this chapter when they study subsequent chapters and those students interested in the economic applications of set theory are referred to *Introduction to Contemporary Microeconomics* by V. C. Walsh.

1.1 Introduction

We are all familiar with the use of the word 'set' in everyday language, but there is no harm in our having a formal definition.

Definition
A *set* is a collection of distinct objects.

For example, we can talk about the set of chairs in a room or the set of people over 65 years of age, or the set of planets in the solar system, or (one of my favourite sets) the set of female students with blue eyes. We shall denote sets by capital letters A, B, S, X, etc.

There are two ways of describing a set; by *enumeration* and by *description*.

If we let A represent the set of days of the week we can write, by *enumeration*
$A = \{$Monday, Tuesday, Wednesday, Thursday, Friday, Saturday, Sunday$\}$ i.e. we list all the members of the set.

Sometimes it may be more convenient to use a *description* of a set.

Example
$A = \{a|a \text{ is a day of the week}\}$
which we read as 'A is the set of all (little) a such that (little) a is a day of the week'.
Notice that the 'common property' that the members of the set share, is included after the vertical bar.

Example
If P is the set of planets in the solar system, we can write, by *enumeration*
$P = \{\text{Mercury, Venus, Earth, Mars, Jupiter, Saturn, Uranus, Neptune, Pluto}\}$
or by *description*
$P = \{p|p \text{ is a planet in the solar system}\}$.

Clearly the more members there are in the set the more advantageous the second method becomes. Indeed, for some sets it may be impossible to use the method of enumeration.

Example
The set $M = \{m|m \text{ is a male living in Great Britain}\}$ could not easily be described by enumeration.

It is convenient at this point to introduce another definition and some more notation.

Definition
An object belonging to a set is called an *element* of the set.

Example
For our previous set
   $M = \{m|m \text{ is a male living in Great Britain}\}$
John Brown is an element of M, which we write as
   John Brown $\in M$
We can also write Jane Brown $\notin M$, i.e. Jane Brown does *not* belong to M.

Before proceeding further with our ideas on sets, it will be useful to introduce the 'real number system' as it is called.

1.2 The real number system

Definitions
The numbers we use for counting, e.g. 1, 2, 3, 4, etc. are called *positive integers*. (They are sometimes called natural numbers.)

The numbers $-1, -2, -3, -4$, etc. are similarly referred to as *negative integers*.

The set of numbers consisting of all the positive integers and the negative integers, with zero included for good measure, is called the set of *integers*.

We use the letter Z to denote the set of integers. So we can write
$Z = \{z | z \text{ is an integer}\}$
and $\quad 2 \in Z$
$\quad -150 \in Z$
$\quad 0 \in Z$

We then have the set of *fractions*, for example $\frac{3}{4}, \frac{7}{8}, -\frac{1}{2}$. These together with the integers form the set of *rational* numbers. We can use the letter Q to represent this set.

So we have $Q = \{q | q \text{ is a rational number}\}$
and $\quad \frac{1}{2} \in Q$
$\quad -\frac{111}{120} \in Q$
$\quad 2 \in Q$

Remark

Rationals are so-called because they are formed by taking the 'ratio' of two integers. Even the number 2 can be thought of as a ratio $\frac{2}{1}$.

Then there are some numbers, like, for example, $\sqrt{2}$ and the number π, which cannot be expressed as one integer over another. These numbers are called the set of *irrational* numbers. There are, in fact, a great many of these irrational numbers but in this book most of the numbers we meet will be from the set of rationals.

If we take all the rational numbers and irrational numbers together to form a set we get the set of *real* numbers. We represent the set of real numbers by the letter R.

So we have $R = \{r | r \text{ is a real number}\}$
$\quad 1 \in R$
$\quad -7 \in R$
$\quad \frac{1}{20} \in R$
$\quad \sqrt{2} \in R$
$\quad 0 \in R$

N.B. It would be impossible to describe any of the sets introduced in this section by enumeration. They contain an *infinite* number of elements, i.e. the list is never-ending.

You might think that the set R contains every number we could ever wish to consider, but there are in fact millions of other numbers, somewhat esoteric in nature, but which do prove useful to mathematicians. These are called *unreal* or *imaginary* numbers. They occur when we try to take the square root of a negative number. Discussion of these is left until Chapter 4.

1.3 More about sets

In this section we shall introduce more definitions which will extend our ideas on sets.

Definition
Two sets A and B are said to be *equal* if they contain *exactly* the same elements.

This may seem a bit pedantic but we shall find we use the concept extensively later in the chapter and it is useful to have the definition formally stated.

Example
If $A = \{1, 2, 4\}$ and $B = \{2, 4, 1\}$
then $A = B$

N.B. The *order* in which the elements are written down does not matter with sets. The set consists of the numbers, *not* of the numbers in a specific order.

Example
If $A = \{a | a$ is a colour of the rainbow$\}$
$B = \{b | b$ is a primary colour$\}$
then $A \neq B$ because, for example, 'orange' is an element of A but not of B. So A and B do not contain exactly the same elements.

Definition
Suppose we have *two* sets X and Y. X is called a *subset* of Y, if every element of X is also an element of Y: i.e. if X is contained in Y, which we write as $X \subset Y$.

Example
If $X = \{1, 2, 3\}$ and $Y = \{1, 2, 3, 4\}$
X is a subset of Y, i.e. $X \subset Y$.

Example
If $A = \{a | a$ is a colour of the rainbow$\}$
 $B = \{b | b$ is a primary colour$\}$
then B is a subset of A, i.e. $B \subset A$.
Example
If $Z = \{z | z$ is an integer$\}$
 $R = \{r | r$ is a real number$\}$
then Z is a subset of R, i.e. $Z \subset R$.

Remark
If there is just *one* element of X which is *not* also an element of Y, then X is *not* a subset of Y. In fact, the easiest way to demonstrate that a set X is *not* a subset of Y is to find such an element.
Example
$X = \{-1, 1, 2\}$
$Y = \{y | y$ is a positive integer$\}$
X is *not* a subset of Y since $-1 \in X$ but $-1 \notin Y$.

Remark
A set is a subset of itself
Example
$Y = \{1, 2, 3, 4\}$
Y is a subset of Y according to our definition, i.e. $Y \subset Y$.

It is convenient to introduce a rather strange set at this point, the *empty set*.

Definition
The empty set is the set which contains *no* elements. It is denoted by the symbol $\emptyset$.

The empty set is considered to be a subset of *every* other set: i.e. $\emptyset \subset A$ for *any* set A.
(If $\emptyset$ were not a subset of any set A, it would be possible to find an element of $\emptyset$ which was not an element of A. However, $\emptyset$ has no elements at all so it is clearly impossible to find such an element.)

Remark
As we shall see later in this chapter, the empty set performs a similar function to that of zero in the real number system. However, we should not confuse $\emptyset$ with $\{0\}$. $\emptyset$ contains no

elements, whereas {0} does contain an element, namely the number 0.

1.4 Operations on sets

We are all familiar with the idea that given any two *numbers* we can combine them to produce a third number using the operations of addition, subtraction, multiplication or division.

For example, given the numbers 5 and 8 we can combine them by addition to produce a third number 13: i.e. $5 + 8 = 13$.
Using a different operation, say multiplication, we get a different number: $5 \times 8 = 40$.

We shall now define operations on *sets*, i.e. ways of combining two sets to produce a third. There are two operations we shall be concerned with – the *union* and the *intersection*.

Definition
Given two sets, A and B, the *union* of these sets, is a set which contains those elements which belong to either A, or B or both.

When combining 5 and 8 by addition we can write $5 + 8$.
Similarly, when combining A and B by taking the union we can write $A \cup B$.

Example
If $A = \{1, 2, 4, 5, 6\}$ and $B = \{1, 3, 4, 9\}$
then $A \cup B = \{1, 2, 3, 4, 5, 6, 9\}$
N.B. If an element appears in both A and B, it is included in $A \cup B$, but only once.

Example
If $M = \{m|m$ is a male human being living in Great Britain$\}$
and $F = \{f|f$ is a female human being living in Great Britain$\}$
then $M \cup F = \{p|p$ is a human being living in Great Britain$\}$.

Example
If A is *any* set then
$A \cup A = A$

Example
If A is *any* set then
$A \cup \emptyset = A$.

The second operation on sets is called *intersection*.

Definition
Given two sets, A and B, the *intersection* of these sets is a set which contains only those elements which belong to both A and B.
We write the intersection of A and B as $A \cap B$. The intersection then contains elements *common* to both sets.
Example
If $A = \{1, 2, 4, 5, 6\}$ and $B = \{1, 3, 4, 9\}$
then $A \cap B = \{1, 4\}$.
Example
For *any* set A
$A \cap A = A$.
Example
For *any* set A
$A \cap \emptyset = \emptyset$
Example
If $M = \{m | m$ is a male human being living in Great Britain$\}$
$F = \{f | f$ is a female human being living in Great Britain$\}$
$M \cap F = \emptyset$, i.e. there is no element which is in M and also in F. There are no common elements.

Definition
When two sets A and B are such that they have no common elements: i.e. $A \cap B = \emptyset$, we say they are *mutually exclusive* or *disjoint*.

We can naturally extend our operations to combine more than two sets. Referring back to our numbers again we can combine 5 and 8 by addition (to give 13) and then combine this with another number (say 4) to produce 17: i.e. $(5 + 8) + 4 = 17$.
The brackets are used in mathematics to tell us that we should combine 5 and 8 before adding 4. (We always work out the expression in the brackets first.) Similarly we have
  $(5 \times 8) \times 4 = 160$.
(Because $5 \times 8 = 40$ and if we then multiply this by 4, we get 160.)

These ideas apply equally well to sets.
Example
If $A = \{1, 2, 4, 5, 6\}$ and $B = \{1, 3, 4, 9\}$
and $C = \{4, 7, 10\}$

then $A \cup B = \{1, 2, 3, 4, 5, 6, 9\}$
and $(A \cup B) \cup C = \{1, 2, 3, 4, 5, 6, 7, 9, 10\}$.
Example
If $A = \{a, b, c\}$ $B = \{b, e\}$ $C = \{d, e, g\}$
then $(A \cup B) = \{a, b, c, e\}$
and $(A \cup B) \cup C = \{a, b, c, d, e, g\}$.
Example
If $A = \{1, 2, 4, 5, 6\}$ $B = \{1, 3, 4, 9\}$ $C = \{4, 7, 10\}$
then $A \cap B = \{1, 4\}$
and $(A \cap B) \cap C = \{4\}$.
Example
If $A = \{a, b, c\}$ $B = \{b, e\}$ $C = \{d, e, g\}$
then $A \cap B = \{b\}$.
$(A \cap B) \cap C = \emptyset$ because $A \cap B$ and C have no elements in common.

We can use our ideas on numbers further. We said $(5 + 8) + 4$ meant combining 5 and 8 first, and then adding 4 to give 17. In fact, if we combine 8 and 4 first, and then add 5 we still get 17: i.e.
    $(5 + 8) + 4 = 17$ and $5 + (8 + 4) = 17$
So $(5 + 8) + 4 = 5 + (8 + 4)$.
So the brackets are not, in fact, needed here, and we just write $5 + 8 + 4 = 17$. Similarly $(5 \times 8) \times 4 = 160$ and $5 \times (8 \times 4) = 160$ so we don't bother with the brackets and we write $5 \times 8 \times 4 = 160$.

If we refer back to our sets we find that this is also true of them: i.e. $(A \cup B) \cup C$ gives exactly the same set as $A \cup (B \cup C)$. (The reader might try this out on the examples above.) So we can ignore the brackets here and write $A \cup B \cup C$. Similarly $(A \cap B) \cap C$ produces the same set as $A \cap (B \cap C)$ and we write $A \cap B \cap C$.
($A \cap B \cap C$ is the set which contains elements common to all three sets.)

The operations on numbers can give us further insight into those on sets.
Suppose we combine 5 and 8 by *addition* to give 13 and then combine this number with 4 by *multiplication* to give 52: i.e. we *mix* our operations.
Then we have $(8 + 5) \times 4 = 52$.

We cannot now dispense with the brackets because $8 + (5 \times 4)$ is not the same: i.e. $(5 \times 4 = 20, 8 + 20 = 28)$.
So when we have an expression where the operations are mixed, the order in which we carry out the operations *does* matter, and we must retain brackets to make it clear what order is required.
Example
$(4 + 3) \times 2 = 14$
whereas $4 + (3 \times 2) = 10$.

Remark
Students may find mathematicians using expressions like $4 + 3 \times 2$, and would be justified in criticising such use. Mathematicians have, over the years, agreed a convention that if any expression contains addition *and* multiplication operations then the multiplications will be done first, unless there are brackets to indicate otherwise.
So $4 + 3 \times 2$ is meant to be taken as $4 + (3 \times 2)$, i.e. 10. If we require the addition to be done first we would have to include the brackets and write $(4 + 3) \times 2$.

In fact, mathematicians have many such conventions and these can cause confusion for students not well versed in the art. If the student is faced with an expression involving mixtures of operations and is unsure which operations to do first, he can resort to a well tried formula concerning the word BEDMAS, which is a mnemonic standing for:
Brackets, **E**xponents, **D**ivisions, **M**ultiplications, **A**ddition, **S**ubtraction, which is the order in which the expression should be evaluated. (Exponents are powers, for example 3^2 or 2^4.)
So an expression like
$(2 + 4) \times 3 - \frac{12}{4} + 2^2$
$= 6 \times 3 - \frac{12}{4} + 2^2$ (**B**rackets first)
$= 6 \times 3 - \frac{12}{4} + 4$ (**E**xponents next)
$= 6 \times 3 - 3 + 4$ (**D**ivision)
$= 18 - 3 + 4$ (**M**ultiplication)
$= 22 - 3$ (**A**ddition)
$= 19$ (**S**ubtraction)

When we return to operations on sets we see a similar pattern.

Example
If $A = \{1, 2, 3, 5\}$ $B = \{1, 6\}$ $C = \{1, 2, 4, 6, 7\}$
then $A \cup B = \{1, 2, 3, 5, 6\}$
and $(A \cup B) \cap C = \{1, 2, 6\}$.
This is not the same as $A \cup (B \cap C)$, since $B \cap C = \{1, 6\}$
and so $A \cup (B \cap C) = \{1, 2, 3, 5, 6\}$.

When the operations in an expression are not all of the same kind, the order of performing the operation is important, and the brackets have to be retained.

Although $(8 + 5) \times 4$ is not the same as $8 + (5 \times 4)$, it is the same as $(8 \times 4) + (5 \times 4)$
since $(8 + 5) \times 4 = 52$
and $(8 \times 4) + (5 \times 4) = 32 + 20 = 52$.
Indeed for *any* three numbers a, b, and c it is true that
$(a + b) \times c = (a \times c) + (b \times c)$.

This property of numbers also extends to sets and we can show that for any three sets
$(A \cup B) \cap C = (A \cap C) \cup (B \cap C)$
and also that
$(A \cap B) \cup C = (A \cup C) \cap (B \cup C)$
although we shall not prove either statement here.

Example
If $A = \{1, 2, 3, 5\}$ $B = \{1, 6\}$ $C = \{1, 2, 4, 6, 7\}$
then $A \cup B = \{1, 2, 3, 5, 6\}$
and so $(A \cup B) \cap C = \{1, 2, 6\}$.
But since $A \cap C = \{1, 2\}$ and $B \cap C = \{1, 6\}$
it is also the case that $(A \cap C) \cup (B \cap C) = \{1, 2, 6\}$.

Recalling our definition of the equality of two sets we can say that $(A \cup B) \cap C = (A \cap C) \cup (B \cap C)$.

1.4.1 Exercises

1. For the sets $A = \{1, 2, 3\}$ $B = \{2, 3, 5\}$ $C = \{4, 5, 7, 9\}$
   find (a) $A \cup B$
   (b) $A \cap B$
   (c) $A \cap C$
   (d) $A \cup B \cup C$
   (e) $A \cup (B \cap C)$
   (f) $A \cap (B \cup C)$

2. For the sets in the previous question, which of the following are true?
   (a) $A \cup (B \cup C) = (A \cup B) \cup C$
   (b) $A \cap (B \cup C) = (A \cap B) \cup C$
   (c) $(A \cap B) \cup C = (A \cup C) \cap (B \cup C)$.
3. If $R = \{r | r$ is a real number$\}$
   $Q = \{q | q$ is a rational number$\}$
   which of the following are true?
   (a) $R \subset Q$
   (b) $Q \subset R$
   (c) $Q \cap R = Q$
   (d) $Q \cup R = Q$
4. If A is *any* set, and B is a subset, what are $A \cup B$ and $A \cap B$?
5. If $W = \{w | w$ is unemployed$\}$
   $X = \{x | x$ is male$\}$
   $Y = \{y | y$ is over 21 years of age$\}$
   $Z = \{z | z$ is over 60 years of age$\}$
   interpret, *in words*, the following:
   (a) $W \cup Z$
   (b) $W \cap X$
   (c) $Y \cap W \cap X$
   (d) John Brown $\in W$
   (e) $X \cap Z \neq \emptyset$
   (f) $Z \subset Y$
   (g) Jane Brown $\notin W$
   (h) $(X \cap W) \cup (X \cap Z)$
   (i) $X \cap (W \cup Z)$.

1.5 Venn diagrams

Venn diagrams (so-called after the mathematician John Venn) are diagrams used to illustrate set operations. A circle is drawn and the area inside the circle is taken to represent the set in question. For example, suppose we are interested in the set A where
$A = \{a | a$ is a female student with blue eyes$\}$.
This set is represented by a circle (Fig. 1.1).

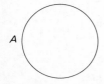

Fig. 1.1

Points inside the circle correspond to elements of A and points outside to elements not in A.

If we now introduce a second set B where
$B = \{b | b$ is a student with blonde hair$\}$
we can represent this set by a similar circle, *on the same diagram*. So we have Fig. 1.2.

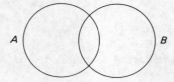

Fig. 1.2

Again, points inside the circle B correspond to elements of the set B and points outside to elements not in B.

Now consider the area shaded //// in Fig. 1.3.

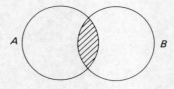

Fig. 1.3

This area lies in both the circle A and the circle B. Points in this area then correspond to elements which are in the set A *and* the set B, i.e. they are in the intersection of A and B. So the shaded area //// corresponds to the set $A \cap B$.

Similarly the set $A \cup B$ is represented on the Venn diagram in Fig. 1.4 by the area shaded \\\\ because any point in this area is in either circle A or circle B (or both) and therefore corresponds to an element which is either in set A or set B (or both), i.e. an element in the set $A \cup B$.

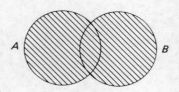

Fig. 1.4

The value of such diagrams becomes evident when we introduce (Fig. 1.5) a third set C where
$C = \{c | c$ is a student under 6 ft tall$\}$.

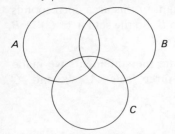

Fig. 1.5

Using ideas similar to those developed above we can represent any set formed by combining A, B and/or C, using $\cup$ or $\cap$, by an area on the Venn diagram.

Example

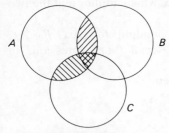

Fig. 1.6

Referring to Fig. 1.6, $A \cap B$ is represented by the area shaded ////, $A \cap C$ is represented by the area shaded \\\\, and the set $(A \cap B) \cup (A \cap C)$ will be represented by the area with *any* shading on (since this area is shaded either //// or \\\\ (or both) and corresponds to elements which are either in $A \cap B$ or $A \cap C$ (or both)).

Example

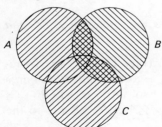

Fig. 1.7

Referring to Fig. 1.7, $A \cup C$ is represented by the area shaded ////, B is represented by the area shaded \\\\, and $(A \cup C) \cap B$ is represented by the area which has both //// shading *and* \\\\ shading, since a point in this area corresponds to an element which is in $A \cup C$ *and* in B. So the set $(A \cup C) \cap B$ is represented by the area shaded ⨯⨯⨯⨯.

We can use Venn diagrams to illustrate the statements we made previously, namely that for any three sets A, B and C
$(A \cup B) \cap C = (A \cap C) \cup (B \cap C)$
and $(A \cap B) \cup C = (A \cup C) \cap (B \cup C)$,
Example
$(A \cup B) \cap C = (A \cap C) \cup (B \cap C)$.
We find the area corresponding to the first set $(A \cup B) \cap C$ on a Venn diagram. We then find the area corresponding to the second set, $(A \cap C) \cup (B \cap C)$, on a separate Venn diagram and observe that the two *areas* are the same, which implies that the two sets are equal.

First let us find the area corresponding to $(A \cup B) \cap C$ (Fig. 1.8). We begin with $A \cup B$. (Recall that expressions in brackets are handled first.)
$A \cup B$ corresponds to the area shaded ////,
C corresponds to the area shaded \\\\,

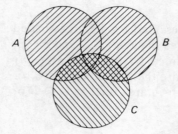

Fig. 1.8

and $(A \cup B) \cap C$ corresponds to the area shaded //// and \\\\, i.e. the area shaded ⨯⨯⨯⨯.

We now look at $(A \cap C) \cup (B \cap C)$ (Fig. 1.9). $A \cap C$ corresponds to the area shaded ////, $B \cap C$ corresponds to the area shaded \\\\, and the union of these two, $(A \cap C) \cup (B \cap C)$ corresponds to the area with shading //// or \\\\, i.e. to the area with any shading at all.

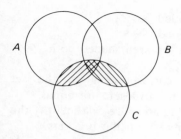

Fig. 1.9

Comparison of this area, corresponding to $(A \cap C) \cup (B \cap C)$, with that corresponding to $(A \cup B) \cap C$ shows the areas are the same, which implies that the sets are the same. That is, $(A \cup B) \cap C = (A \cap C) \cup (B \cap C)$

Example

$(A \cap B) \cup C = (A \cup C) \cap (B \cup C)$.

First we find the area corresponding to $(A \cap B) \cup C$ (Fig. 1.10).

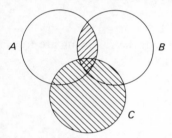

Fig. 1.10

$A \cap B$ corresponds to the area shaded ////.
C corresponds to the area shaded \\\\.
$(A \cap B) \cup C$ then corresponds to the area with *any* shading on, i.e. shaded either //// or \\\\.

Then we find the area corresponding to $(A \cup C) \cap (B \cup C)$ (Fig. 1.11).

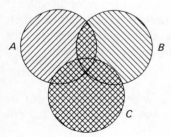

Fig. 1.11

$A \cup C$ corresponds to the area shaded ////,
$B \cup C$ corresponds to the area shaded \\\\,
$(A \cup C) \cap (B \cup C)$ corresponds to the area shaded both //// and \\\\, i.e. the area shaded ✕✕✕✕.
This is the same area as that corresponding to the set $(A \cap B) \cup C$ and this implies that the two sets are equal.

N.B. Notice how when identifying the areas we first identify the areas corresponding to the sets within the brackets.

1.5.1 Exercises

Given *any* three sets A, B and C,
1. Demonstrate the validity of the following laws concerning set operations, using Venn diagrams
   (a) $A \cup (B \cap C) = (A \cup B) \cap (A \cup C)$
   (b) $A \cap (B \cap C) = (A \cap B) \cap C$
   (c) $A \cup (B \cup C) = (A \cup B) \cup C$
   (d) $A \cap (B \cup C) = (A \cap B) \cup (A \cap C)$.
2. Using Venn diagrams demonstrate that the following are true:
   (a) $(A \cup B) \cap C \neq A \cup (B \cap C)$
   (b) $(A \cap B) \cup C \neq A \cap (B \cup C)$.

1.6 Complements

Definition

It is often useful when discussing sets in a particular application to consider one set which contains every possible element (and set) we might wish to talk about in that application. This set we call the *universal set*, and we use the letter U to denote it.

For example, if we were discussing the way people voted in an election we might take as our universal set, the set of all voters.

If we were discussing numbers, our universal set might be taken as the set of real numbers or just the set of integers if we were sure that we would not be considering any non-integer values.

Definition

We can now define the *complement of a set* A to be the set of elements which do not belong to A, but do belong to U. We denote the complement by $\bar{A}$.

Example
If $U = \{u | u$ is a voter$\}$
$A = \{a | a$ votes for a particular party$\}$
then $\bar{A}$ is the set of voters who do *not* vote for that party.
Example
$U = \{u | u$ is an integer$\}$
$E = \{e | e$ is an even integer$\}$
then $\bar{E}$ is the set of odd integers.

On a Venn diagram we can represent the universal set by a rectangle and $\bar{A}$ by the area *outside* the circle representing A (Fig. 1.12).

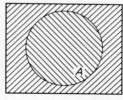
Fig. 1.12

We can extend our definition to include sets like $\overline{A \cup B}$, which is the complement of $A \cup B$, and $\overline{A \cap B}$ which is the complement of $A \cap B$.

We can also have $\bar{A} \cup \bar{B}$ which is the *union* of $\bar{A}$ and $\bar{B}$ (this is not the same as $\overline{A \cup B}$), and $\bar{A} \cap \bar{B}$.

These sets are related as follows:
$$\overline{A \cup B} = \bar{A} \cap \bar{B}$$
and $\overline{A \cap B} = \bar{A} \cup \bar{B}$.
(Notice that $\overline{A \cup B} \neq \bar{A} \cup \bar{B}$ in general.)

We can demonstrate these relationships using Venn diagrams.
Example
$\overline{A \cup B} = \bar{A} \cap \bar{B}$.
First consider $\overline{A \cup B}$ (Fig. 1.13).

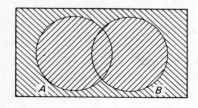
Fig. 1.13

$A \cup B$ is shaded ////, so $\overline{A \cup B}$ is represented by the area shaded \\\\.

Now consider $\bar{A} \cap \bar{B}$ (Fig. 1.14).
$\bar{A}$ is the area shaded ////,
and $\bar{B}$ is the area shaded \\\\.

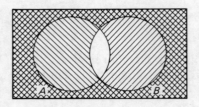

Fig. 1.14

Consequently $\bar{A} \cap \bar{B}$ is the area shaded both //// and \\\\, i.e. the area shaded ✕✕✕✕,
which is the same as the area representing $\overline{A \cup B}$.

1.6.1 Exercises

1. If $U = \{u | u$ is an integer$\}$
   $A = \{a | a$ is an integer less than 2$\}$
   $B = \{b | b$ is an integer greater than 4$\}$.
   What are the following sets?
   (a) $\bar{A}$
   (b) $\bar{B}$
   (c) $\overline{A \cup B}$
   (d) $\bar{A} \cup \bar{B}$
   (e) $\overline{A \cup B}$
   (f) $\bar{A} \cap \bar{B}$
2. Use Venn diagrams to show that:
   (a) $\overline{A \cap B} = \bar{A} \cup \bar{B}$.
   (b) $\overline{A \cap B} \neq \bar{A} \cap \bar{B}$.

1.7 Applications to economics

To build up the necessary background to enable us to analyse meaningful economic problems using set theory would require more space than is available in this section, and we shall not, therefore, attempt it. The reader will discover that most of the mathematics, and consequently the applications to economics, in subsequent chapters, are built on the foundation of sets and this in itself justifies their inclusion in this book. However, set theory is useful in its own right for analysing economic problems and provides a framework which is in many ways less restrictive than the conventional mathematics described and utilised in subsequent chapters of this book.

The reader interested in pursuing this role of sets in economic theory is recommended to read *Introduction to Contemporary Microeconomics* by V. C. Walsh, which provides not only an entertaining and relatively easy introduction to the subject, but also a discussion of the limitations of the approach to economic problems presented in the remainder of this book.

Chapter 2

Mappings and functions

Economics is concerned with relationships between different economic quantities and the effects that changes in one economic quantity might have on another. These relationships are defined by *functions*, and in this chapter we define mappings (or relations) and functions, as well as discussing the use of graphs to represent such functions. In section 2.10 we then apply the definitions to economics, by defining demand and supply functions for a simple market model, and cost, revenue and profit functions for a firm producing one commodity. These functions are used extensively in the remaining chapters of the book.

2.1 Introduction

Suppose we have two sets
$A = \{a | a \text{ is a person}\}$
$B = \{b | b \text{ is a colour}\}$.
These sets appear to have very little in common, for instance $A \cap B = \emptyset$.

However, if we were discussing the colour of a person's eyes, they would be connected. Every element in set A can be matched to an element in set B. For example, John Brown can be matched to the colour blue, if it happens that John Brown has blue eyes. We write this as
John Brown → blue.
Jane Brown has green eyes and so we can write
Jane Brown → green.

We say the first set A is *mapped* to the second set B. We have a *mapping* between the sets, and we write
$A \to B$.

therefore, attempt it. The reader will discover that most of the mathematics, and consequently the applications to economics, in subsequent chapters, are built on the foundation of sets and this in itself justifies their inclusion in this book. However, set theory is useful in its own right for analysing economic problems and provides a framework which is in many ways less restrictive than the conventional mathematics described and utilised in subsequent chapters of this book.

The reader interested in pursuing this role of sets in economic theory is recommended to read *Introduction to Contemporary Microeconomics* by V. C. Walsh, which provides not only an entertaining and relatively easy introduction to the subject, but also a discussion of the limitations of the approach to economic problems presented in the remainder of this book.

Chapter 2

Mappings and functions

Economics is concerned with relationships between different economic quantities and the effects that changes in one economic quantity might have on another. These relationships are defined by *functions*, and in this chapter we define mappings (or relations) and functions, as well as discussing the use of graphs to represent such functions. In section 2.10 we then apply the definitions to economics, by defining demand and supply functions for a simple market model, and cost, revenue and profit functions for a firm producing one commodity. These functions are used extensively in the remaining chapters of the book.

2.1 Introduction

Suppose we have two sets
 $A = \{a | a \text{ is a person}\}$
 $B = \{b | b \text{ is a colour}\}$.
These sets appear to have very little in common, for instance $A \cap B = \emptyset$.

However, if we were discussing the colour of a person's eyes, they would be connected. Every element in set A can be matched to an element in set B. For example, John Brown can be matched to the colour blue, if it happens that John Brown has blue eyes. We write this as
 John Brown $\to$ blue.
Jane Brown has green eyes and so we can write
 Jane Brown $\to$ green.

We say the first set A is *mapped* to the second set B. We have a *mapping* between the sets, and we write
 $A \to B$.

Example
If $A = \{a \mid a$ is a student$\}$
   $B = \{b \mid b$ is a real number$\}$,
then set A can be *mapped* to set B: i.e. $A \rightarrow B$, if we are discussing the height, in centimetres, of students.

Let us have a formal definition of a *mapping*.

Definition
A *mapping* consists of two sets and a rule for assigning to each (and every) element of the first set, one (or more) elements of the second set.

Example
If $Z = \{z \mid z$ is an integer$\}$
   $R = \{r \mid r$ is a real number$\}$
then we can define a mapping from Z to R, $Z \rightarrow R$ using the rule 'halve the integer'.
So, for example, $\quad 2 \rightarrow 1$
$\qquad\qquad\qquad\quad 3 \rightarrow 1.5$
$\qquad\qquad\quad -4 \rightarrow -2.$

Notice that *every* element of the first set, which we shall call the *domain*, is matched to at least one element in the second set, which is called the *codomain*, but not every element in the second set (the codomain) is associated with an element in the first set (the domain).

In the above example there is no integer which gets mapped to the number 1.3 in the codomain.

Example
There are often situations where it is useful to use the same set for both the domain and codomain.
If $\; R = \{r \mid r$ is a real number$\}$
we can define a mapping $R \rightarrow R$ according to the rule 'double the number'.
So, for example, $\quad 2 \rightarrow 4$
$\qquad\qquad\qquad\quad 1.5 \rightarrow 3$
$\qquad\qquad\quad -2.39 \rightarrow -4.78.$

Notation
To describe the mapping from the set of people to the set of colours according to the colour of a person's eyes, we can give the

mapping a name, say m, and write
m: Set of people → set of colours.
Similarly, for an element of the domain
m: John Brown → blue
which can also be written as
m(John Brown) = blue.

2.2 The image set

Definition
If $a \in A$ and $b \in B$ and if $m: a \to b$
then we say that b is the image of a.
Example
m: Fred Smith → blue
and we say blue is the image of Fred Smith under this mapping.
Example
If $R = \{r | r \text{ is a real number}\}$
and $f: R \to R$ according to the rule 'double the number',
then $f: 3 \to 6$ and we say 6 is the image of 3 under the mapping.
Similarly, the image of -13 is -26.

We have already pointed out that while *every* element in the domain is associated with at least one element in the codomain, *not every* element in the codomain has an associated element in the domain.

In other words, every element in the domain has an image but not every element in the codomain is the image of an element in the domain.

Definition
Those elements in the codomain which are images clearly form a subset of the codomain and we call this subset the *image set*.
Example
If $A = \{1, 2, 3\}$
  $B = \{1, 2, 3, 4, 5, 6, 7\}$
and $m: A \to B$ according to the rule 'double the number',
we have $1 \to 2$
          $2 \to 4$
          $3 \to 6$
i.e. the image set is $\{2, 4, 6\}$ which is a subset of the codomain.

Example
If $R = \{r | r$ is a real number$\}$
and $f: R \to R$ according to the rule 'square the number',
then $f: 2 \to 4$
$f: -3 \to 9$
$f: 1.3 \to 1.69.$

The image set is the set of *positive* reals (with zero) because any number when squared becomes positive, i.e. only positive real numbers (or zero) are images.

Example
Sometimes the image set can be the whole of the codomain.
If $R = \{r | r$ is a real number$\}$
and $g: R \to R$ according to the rule 'add 2 to the number', then the image set is the set R.

2.2.1 Exercises

1. If $A = \{a | a$ is a person$\}$
   and $Z = \{z | z$ is an integer$\}$
   and $m: A \to Z$ such that each person is associated with the date of the month of his/her birthday, then what is
   (a) The domain of the mapping?
   (b) The codomain?
   (c) The image of you?
   (d) The image set?

2. If $R = \{r | r$ is a real number$\}$
   and $f: R \to R$ according to the rule 'square the number and subtract 1', then what
   (a) Are the images of 2, 0, −5?
   (b) Is the image set of the mapping?

2.3 Types of mappings

Consider the following four mappings:
(a) m: People $\to$ real numbers, according to a person's height in centimetres today.
(b) n: Men with daughters $\to$ people, such that a man is mapped to his daughter.
(c) p: Parents $\to$ people, such that a person is mapped to his or her daughter.

(d) q: Countries → cities, such that a country is mapped to its *capital city*.

For mapping m, each element of the domain is associated with one, and only one element in the codomain. (A person can only have one height.)

However, it is possible for more than one element in the domain to be associated with the same element in the codomain. (It is possible to find two people with the same height.) We say mappings of this kind are *'many to one'*.
This type of mapping is illustrated in Fig. 2.1.

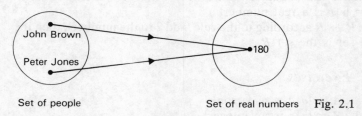

Set of people   Set of real numbers   Fig. 2.1

In this figure we see that a typical element of the domain John Brown, is mapped to the element 180 in the codomain; he cannot also be mapped to another element in the codomain. However, Peter Jones is also mapped to 180 and we have a 'many to one' mapping.

Consider now mapping n. An element in the domain, e.g. John Brown, can now be mapped to more than one element in the codomain if he has more than one daughter. However, it is not possible for two elements in the domain to be mapped to the same element in the codomain. (No two men can have the same daughter.)

We call this kind of mapping a 'one to many' mapping, and is illustrated in Fig. 2.2.

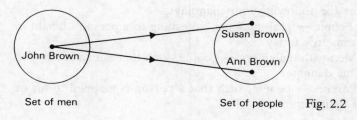

Set of men   Set of people   Fig. 2.2

In our example, John Brown has two daughters, Susan and Ann.

Mapping *p* is of a type known as 'many to many'. Here it is possible for two elements in the domain to be mapped to the same element in the codomain. (Both John Brown and Janet Brown, his wife, are mapped to Susan and indeed to Ann.) The mapping is both 'many to one' *and* 'one to many' (Fig. 2.3).

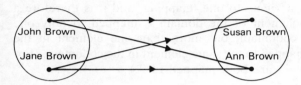

Set of parents   Set of people   Fig. 2.3

The final type of mapping is illustrated by mapping *q* (Fig. 2.4). This is known as a 'one to one' mapping. Every element in the domain is mapped to *only one* element in the codomain (no country has more than one capital city), and no two elements in the domain are mapped to the same element in the codomain. (No two countries have the same capital city.)

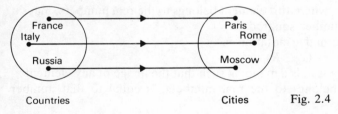

Countries   Cities   Fig. 2.4

The mappings which are of interest to economists are usually of the 'one to one' or 'many to one' kind. These mappings are usually referred to as functions.

Definition
A *function* is a mapping for which each element in the domain has only one element in the codomain associated with it.

We can see from our definition that a function is either a many to one or a one to one mapping.

We shall be primarily interested in the rest of the book in mappings, and particularly functions, between the real numbers.
Example
$R = \{r | r \text{ is a real number}\}$
$m: R \to R$ according to the rule 'square the number'.
So, for example, $m: 2 \to 4$
$m: 3 \to 9$
$m: -3 \to 9.$

This mapping is a 'many to one' mapping, and it is therefore a function. Each element of the domain is mapped to only one element in the codomain, but more than one element in the domain can be mapped to the same element in the codomain. (For example, 3 and -3 are both mapped to 9.)

2.4 Notation

When describing mappings between the real numbers it is convenient to use a special notation. So we might describe the mapping m above, by writing
$m: x \to x^2 \ (x \in R)$
which simply means m is a mapping which, when applied to an element x, where this element belongs to the real numbers, maps it to that number squared.

An alternative notation is
$m(x) = x^2 \ (x \in R)$
which means m is a mapping such that the image of any element x, where x belongs to the real numbers, is equal to that number squared.

Let us illustrate this notation with some more examples. In all the examples $R = \{r | r \text{ is a real number}\}$.
Example
$n: x \to 2x \ (x \in R)$
This mapping n, maps any element of the real numbers (i.e. any real number) to twice that number.
So $n: 1 \to 2$
$n: 2 \to 4$
$n: -3 \to -6.$

This mapping is a 'one to one' mapping. Each element of the

domain is mapped to only one element in the codomain and no two elements in the domain are mapped to the same element in the codomain. (Can you find two?) It is therefore a function.

Example
$f(x) = 1 - x \ (x \in R)$.
So, for example,
$f : 1 \to 0$ or $f(1) = 0$
$f : 2 \to -1$ or $f(2) = -1$
$f : 5 \to -4$ or $f(5) = -4$
$f : -3 \to 4$ or $f(-3) = 4$ (because $1 - (-3) = 1 + 3 = 4$).
This mapping is also 'one to one' and, hence, is a function.

Example
$g : x \to \sqrt{x} \ (x \in R)$
So, for example, $g : 4 \to 2$.
But also $g : 4 \to -2$ because both 2 and -2 when squared give 4. So the square root of 4 can be either $+2$ or -2.

This mapping is 'one to many' and not a function.

Remark
Another notation frequently used, and one with which you may be familiar is to use the letter y to represent the image of an element in the domain.
So, for example, the mapping
  $f : x \to x^2 \ (x \in R)$
which we have said can be written
  $f(x) = x^2 \ (x \in R)$
under the 'y notation' would be written
  $y = x^2 \ (x \in R)$.

Example
The mapping $y = 3x + 2 \ (x \in R)$
is the same as $f(x) = 3x + 2 \ (x \in R)$
or $f : x \to 3x + 2 \ (x \in R)$

Example
The mapping $y = 1 - x \ (x \in R)$
is the same as $f(x) = 1 - x \ (x \in R)$
or $f : x \to 1 - x \ (x \in R)$.

In this book we shall mainly use the '$f(x)$' notation. Since nearly all our mappings have the domain as the set of real numbers we shall omit the $(x \in R)$ from the description. The set of real numbers will be assumed to be the domain unless otherwise stated. Also we shall refer to the mappings as functions when appropriate.

So we shall describe our mappings (or in most cases they will be functions) as, for example,
$f(x) = 2x - 5$.

2.4.1 Exercises

1. For the mapping $m(x) = 4x + 2$
   (a) Find the images of 0, 2, −1, 4 under the mapping.
   (b) What type of mapping do you think m is? Is it a function?
2. For the mapping $f(x) = 2x^2 - 1$
   (a) Find the images of 1, 3, −1, under the mapping.
   (b) What type of mapping is it? Is it a function?

2.5 Graphs

Given any mapping we can, for each element in the domain, find one, or more, associated elements in the codomain.

Example
For the function $f(x) = x^2$
$f: 3 \rightarrow 9$ or $f(3) = 9$
$f: -2 \rightarrow 4$ or $f(-2) = 4$
$f: 1 \rightarrow 1$ or $f(1) = 1$

Definition
A pair, e.g. (3, 9), where the first member is an element in the domain, and the second the associated element in the codomain is called an *ordered pair*. (Ordered because it is important that the *first* member be from the *domain* and the second from the *codomain*.)

For the function $f(x) = x^2$ other ordered pairs are
  (−2, 4) (because −2, an element in the domain, is mapped to 4 in the codomain)
  (1, 1)
  (−5, 25)
  (2, 4), etc.

Example
$m(x) = 3x - 1$ has ordered pairs
(1, 2) (because 1 is mapped to 2)
(2, 5)
(0, −1)
(−1, −4), etc.

Remark
It is often the practice to write these ordered pairs in the form of a table.
Example
For $m(x) = 3x - 1$ the ordered pairs can be written as

x	$m(x)$
1	2
2	5
0	−1
−1	−4

These ordered pairs of the function form the basis for drawing the *graph* of the function.

N. B. Our example was a function but ordered pairs can be obtained for any type of mapping.

Let us take the function $m(x) = 3x - 1$ as an example to illustrate the idea of a graph.

The domain of the function is the set of real numbers and this set can be represented by a *horizontal straight line*, with different points on the line corresponding to different elements in the set.

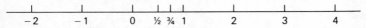

Any element in the domain will correspond to some point on this line (often referred to as *the real line*).

Since the codomain of our function is also the set of real numbers, this can also be represented by a line, and elements in the codomain correspond to points on the line. For the purpose of drawing graphs, it is convenient to represent the codomain by a *vertical line*, on the same diagram as the horizontal line representing the domain, with the points representing zero for both sets coinciding (Fig. 2.5).

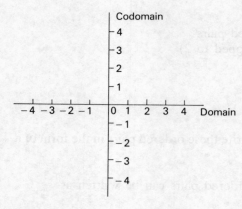

Fig. 2.5

Remark

Often the horizontal line is labelled x, to represent elements from the domain, and the vertical line labelled $f(x)$, to represent the elements from the codomain.

Now any ordered pair can be represented by a point on the diagram. For example, the ordered pair (1, 2) will correspond to the point A on the diagram (Fig. 2.6). (The point is opposite the 1 on the horizontal line, corresponding to the domain, and opposite 2 on the vertical line, corresponding to the codomain.) Notice that the ordering of the pair is important. The ordered pair (2, 1) corresponds to the point B on the diagram because the 2 is now from the domain (and hence opposite the 2 on the horizontal line) and the 1 is from the codomain (and hence opposite the 1 on the vertical line).

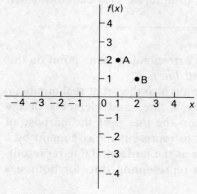

Fig. 2.6

Let us now return to our example $m(x) = 3x - 1$. Some ordered pairs for this are:

x	$m(x)$
1	2
2	5
0	−1
−1	−4

If we now represent *each* of these ordered pairs on a diagram we get Fig. 2.7.

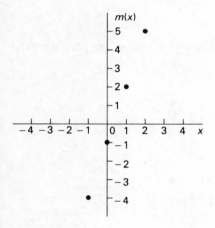

Fig. 2.7

If we considered *every* element in the domain (and there are an infinite number of them), and placed the corresponding ordered pairs on the diagram, we would, in fact, end up with a line (Fig. 2.8). This line is called the *graph of the mapping or function*. The graph is really made up of an infinite number of points, each one of which corresponds to an ordered pair of the function.

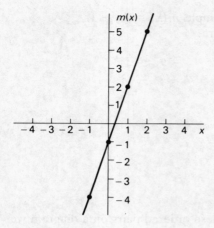

Fig. 2.8

Different mappings produce different ordered pairs and hence different graphs.

Example

$f(x) = x^2$.

Some ordered pairs are given below and if these are placed on the diagram we get Fig. 2.9.

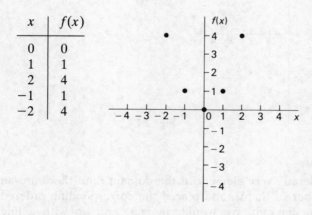

x	$f(x)$
0	0
1	1
2	4
−1	1
−2	4

Fig. 2.9

If we were to place *all* the ordered pairs of the function on the diagram our graph would be as in Fig. 2.10.

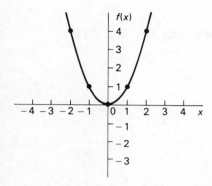

Fig. 2.10

which we can see is a different graph from that for the function $m(x) = 3x - 1$.

N. B. When drawing graphs we cannot, of course, calculate *all* the ordered pairs and represent them on the diagram. (There are an infinite number of ordered pairs.) We calculate a few ordered pairs, represent these on the diagram and then 'guess' at the likely shape of the graph. In section 2.7 we shall find that the guessing can be made more reliable.

2.5.1 Exercises

Draw the graphs of the following functions:

(1) $f(x) = 2x - 1$

(2) $g(x) = 1 - 2x$

(3) $h(x) = 1 + x^2$

(4) $l(x) = \dfrac{1}{x}$ $\qquad (x \neq 0)$

(5) $m(x) = 2x^2 + x - 3$.

2.6 More on graphs

We have seen how, given a mapping, we can produce the ordered pairs of the mapping and hence the graph of the mapping. Alternatively, if we are given the graph of a mapping we can find the ordered pairs from the graph.

Example
Suppose the graph of a function looks like Fig. 2.11.

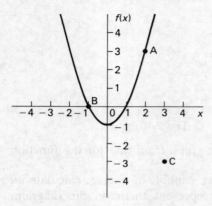

Fig. 2.11

Remember that every point on the graph corresponds to an ordered pair of the function. So for example the point A, which is on the graph, and opposite 2 on the horizontal line (corresponding to the domain) and 3 on the vertical line (corresponding to the codomain) represents the ordered pair (2, 3). Therefore the pair (2, 3) is an ordered pair of the function, i.e. the function maps 2 to 3.

$f : 2 \to 3$.

Similarly using point B, opposite -1 on the horizontal and 0 on the vertical, we know that $f : -1 \to 0$.

The fact that the point C, representing the ordered pair $(3, -3)$, is *not* on the graph tells us that f does *not* map $+3$ from the domain to -3 in the codomain.

So we can, from the graph find *all* the ordered pairs of the function by considering *all* the points on the graph. (The function is in fact given by the formula $f(x) = x^2 - 1$, but although we can find all the ordered pairs from the graph, it would not usually be possible to find the formula.)

Remark
We have concerned ourselves only with drawing graphs of function but there is no reason why we cannot draw the graphs of mappings between the real numbers which are not functions.

Example
$m(x) = \sqrt{x}$ (where x is a number greater than or equal to zero)

has ordered pairs

x	$m(x)$
0	0
1	+1
1	−1
2	+1.414
2	−1.414

(Remember $\sqrt{x}$ can be either positive or negative.)

The graph of this mapping (which is clearly *not* a function, since an element such as 2 is mapped to more than one element in the codomain) is shown in Fig. 2.12.

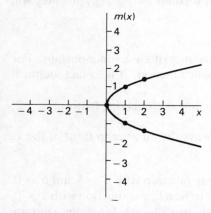

Fig. 2.12

(Notice how the graph is to the right of the vertical axis. This is because the domain contains no negative numbers.)

2.7 Named functions

Some functions are more frequently used than others and are therefore given names so that they can be recognised more easily. In this section we shall consider four such names – linear, constant, quadratic, polynomial. Later we shall meet others.

2.7.1 Linear functions

We saw in section 2.5 how the graph of the function $m(x) = 3x - 1$ turned out to be a straight line.

Definition
In fact *any* function of the form $f(x) = ax + b$ where a and b are any two real numbers (positive or negative) turns out to have a graph which is a straight line.

Consequently *all* such functions are referred to as *linear functions*. So, for *example*, the following are linear functions:

$f(x) = 4x + 7 \quad (a = 4, b = 7)$
$f(x) = x + 2 \quad (a = 1, b = 2)$
$f(x) = 2x - 3 \quad (a = 2, b = -3)$
$f(x) = \tfrac{1}{2}x - 1 \quad (a = \tfrac{1}{2}, b = -1)$

They are all linear functions and if you draw their graphs they will be straight lines.

Remark
The order in which the formula is written is unimportant. For example, $f(x) = 7 + 4x$ is a linear function. It is in fact identical to $f(x) = 4x + 7$.
Similarly, $f(x) = 3 - 2x$ is a linear function (with $a = -2$, $b = 3$).
Notice that the 'a' number is always the number in front of the x.

Remark
The function $f(x) = 3x$ is a linear function with $a = 3$ and $b = 0$. We could also consider $f(x) = 4$ to be a linear function with $a = 0$, $b = 4$ but it is more convenient to call such functions *constant functions* (see section 2.7.2).

Remark
When the graph of a linear function $f(x) = ax + b$ is drawn, it turns out that the 'a' number gives the 'slope' of the line and the 'b' number the point on the vertical axis where the line crosses that axis (often called the 'intercept').

Example
$f(x) = 2x + 4$ with $a = 2$ and $b = 4$ has the graph in Fig. 2.13. It has a slope equal to 2 (i.e. for every one unit you move to the right you go up 2 units) and crosses the vertical axis at the point 4.

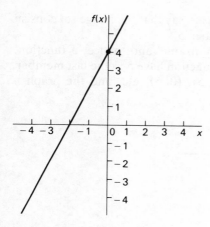

Fig. 2.13

Remark
All linear functions are 'one to one' (except for constant functions).

2.7.2 Constant functions

Definition
We have already introduced in section 2.7.1 the idea of a *constant function* as a function of the form $f(x) = b$ for some real number b.

Example
$f(x) = 1$
$f(x) = 5$
$f(x) = -3$
are all constant functions.

Let us look more closely at these functions by considering, for example,
   $f(x) = 5$.
To understand this function, consider the following:
x represents *any* element in the domain. The image of *any* x then, under the function f, is equal to 5, i.e. every element in the domain is mapped to the same element, 5, in the codomain.
   $2 \to 5$
   $1 \to 5$
   $-3 \to 5$.

37

Another way of expressing this is to say that the image set consists of just one element – the number 5.

The mapping is clearly 'many to one' and, hence, a function.

All the ordered pairs for the function have 5 as the last member, e.g. (2, 5), (1, 5), (−3, 5), (7, 5), (0, 5), etc. and the graph is shown in Fig. 2.14.

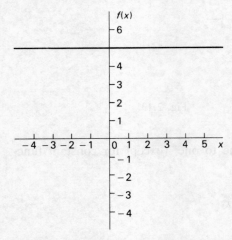

Fig. 2.14

So the graph of a constant function $f(x) = b$ is a straight horizontal line through the point b on the vertical axis. Viewed as a linear function the a number is 0, i.e. the graph has zero slope.

2.7.3 Quadratic functions

Definition
A function of the form $f(x) = ax^2 + bx + c$, where a, b and c, are any real numbers, is called a *quadratic function*.

Example
$f(x) = 3x^2 + 4x + 7$
is a quadratic function (with $a = 3, b = 4, c = 7$).
So are $f(x) = 2x^2 - 3x - 1$ ($a = 2, b = -3, c = -1$)
and $f(x) = x^2 + 2x + 1$ ($a = 1, b = 2, c = 1$).

Remark
The order of the formula is unimportant.

Example
$f(x) = 1 + 2x + x^2$ is a quadratic function (it is the same as $f(x) = x^2 + 2x + 1$).
However, it is important (as we shall see later) to ensure that we denote the number on x^2 as a, that on x as b and the remaining number as c.
Example
$f(x) = 7 - 3x + 2x^2$ is a quadratic function with $a = 2$, $b = -3$, $c = 7$.

Remark
The b value and the c value can be 0, but conventionally we consider that only functions with $a \neq 0$ should be called quadratic.
Example
$f(x) = 3x^2 + 2$ is a quadratic ($a = 3$, $b = 0$, $c = 2$),
$f(x) = x^2 + 7x$ is a quadratic ($a = 1$, $b = 7$, $c = 0$),
and even $f(x) = 3x^2$ is a quadratic ($a = 3$, $b = 0$, $c = 0$).
But if, for example, $a = 0$, $b = 2$, $c = 7$, i.e. $f(x) = 2x + 7$, it is *not* called a quadratic. (It is, of course, a linear function.)

Remark
All quadratic functions are 'many to one'.

Remark
If we draw the graph of, for example, $f(x) = x^2 + 2x + 1$, we get Fig. 2.15.

x	$f(x)$
0	1
1	4
-1	0
-2	1

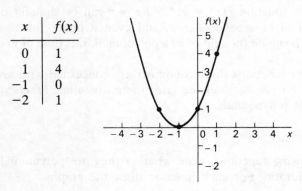

Fig. 2.15

In fact, all quadratic functions have graphs which are either U-shaped or inverted U-shaped (Fig. 2.16),
i.e.

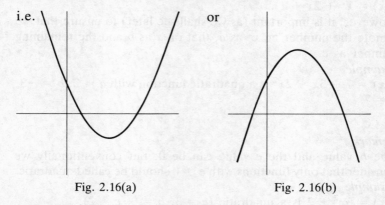

Fig. 2.16(a) Fig. 2.16(b)

2.7.4 Polynomial functions

Astute students, having seen the linear function $f(x) = ax + b$, and the quadratic $f(x) = ax^2 + bx + c$, may well be wondering if we have names for functions like $f(x) = ax^3 + bx^2 + cx + d$ or even $f(x) = ax^4 + bx^3 + cx^2 + dx + e$.

In fact these functions are called *polynomials*.

A function like $f(x) = ax^3 + bx^2 + cx + d$ is a *third degree polynomial*.

Functions of the form $f(x) = ax^4 + bx^3 + cx^2 + dx + e$ are *fourth degree polynomials*.

Functions of the form $f(x) = ax^5 + bx^4 + cx^3 + dx^2 + ex + f$ are *fifth degree polynomials*, and so on.

A quadratic function $f(x) = ax^2 + bx + c$ can be thought of as a polynomial of second degree, and even a linear function $f(x) = ax + b$ can be thought of as a polynomial, this time of first degree.

Many of the functions that economists are concerned with are polynomials, but as we shall see later there are other functions which are not polynomials.

2.7.5 Exercises

For the following functions, state whether they are polynomial, linear or quadratic. For each function draw the graph.

1. $f(x) = 4x - 2$
2. $f(x) = 3x^2 + 2x + 1$
3. $f(x) = 1 - x$
4. $f(x) = 1 - 3x + x^2$
5. $f(x) = \dfrac{1}{x} \qquad (x \neq 0)$
6. $f(x) = x^3 + 3x^2 + x - 1.$

2.8 Reverse mappings and inverse functions

Consider the following mapping from the set $A = \{a \mid a \text{ is a parent}\}$ to the set $B = \{b \mid b \text{ is a person}\}$.
$m: A \to B$ such that each parent is mapped to his or her children.
Example
  m: John Brown $\to$ Susan Brown
  m: John Brown $\to$ Ann Brown
  m: Jane Brown $\to$ Susan Brown
  m: Jane Brown $\to$ Ann Brown
(Recall that John and Jane Brown have two daughters, Susan and Ann.)

Definition
It is possible to define the *reverse mapping* m^* such that $m^*: B \to A$ according to the rule that a person is mapped to his or her parents.
So, for example, m^*: Susan Brown $\to$ John Brown.
Diagrammatically we have Fig. 2.17.

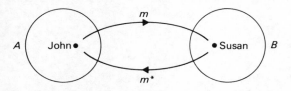

Fig. 2.17

Notice that our mapping m is 'many to many' as indeed is our mapping m^*.

If a mapping is 'many to one' (i.e. a function) our reverse mapping will be 'one to many', for example
if $A = \{a \mid a \text{ is a person}\}$
$R = \{r \mid r \text{ is the set of real numbers}\}$
and $m: A \rightarrow R$ according to the rule that a person is given his or her height (in cms) then, for example
m: John Brown $\rightarrow$ 180
m: Peter Jones $\rightarrow$ 180
(because both John and Peter have the same height, namely, 180 cms). Clearly m is 'many to one'.
$m^*: R \rightarrow A$ will match for example
$m^*: 180 \rightarrow$ John Brown
$m^*: 180 \rightarrow$ Peter Jones
i.e. m^* is 'one to many' – not a function.

Definition
When the original mapping m is a 'one to one' function the reverse mapping will also be a 'one to one' function. In this case the reverse mapping is called the *inverse function* and labelled m^{-1}.
Example
$f(x) = 3x^2$
This is a quadratic function and hence 'many to one'.
Its reverse mapping is

$$f^*(x) = \sqrt{\frac{x}{3}}$$

and is 'one to many' – not a function (we can take either the positive or negative value for the square root).
Example
$f(x) = 2x - 3$
This is a linear function and hence 'one to one'.
Its *inverse function* is

$$f^{-1}(x) = \frac{x + 3}{2}$$

and, for example,

$$1 \xrightarrow{f} -1 \xrightarrow{f^{-1}} 1$$
and $\quad 4 \xrightarrow{f} 5 \xrightarrow{f^{-1}} 4.$

2.9 Some more named functions

2.9.1 Exponents

We have already introduced expressions like x^2 in our quadratic functions, and most readers will not have been unduly worried by this. However, a brief description of the use and manipulation of exponents will not go amiss.

We define exponents (or powers or indices) as follows:

Definition
Given any number x, and any *positive integer n*,
then $x^n = \underbrace{x \cdot x \cdot x \cdots x}_{n \text{ times}}$

(n is known as the exponent, or power, or index, of x)
Example
$x^4 = x \cdot x \cdot x \cdot x$
and if $x = 3$ then $3^4 = 3 \cdot 3 \cdot 3 \cdot 3 = 81$.

Remark
$x^1 = x$

Remark
$f(x) = x^4$, and indeed $f(x) = x^n$ for any positive integer n, defines a function.

The above definition is only valid providing n is a positive integer. If n is a fraction or negative or even zero we have the following three definitions.

Definition
Given any number x and any positive integer n then
   $x^{1/n} = \sqrt[n]{x}$, i.e. the 'n^{th} root' of x.
Example
$x^{1/2} = \sqrt{x}$
and if $x = 9$ then $9^{1/2} = \sqrt{9} = 3$.
Example
$x^{1/3} = \sqrt[3]{x}$
and if $x = 8$ then $8^{1/3} = \sqrt[3]{8} = 2$.

Definition
Given any number x and any positive integer or fraction n then
$$x^{-n} = \frac{1}{x^n}.$$
Example
$$x^{-2} = \frac{1}{x^2}$$
and if $x = 3$ then $3^{-2} = \frac{1}{3^2} = \frac{1}{9}$.

Example
$$x^{-1} = \frac{1}{x^1} = \frac{1}{x}$$
and if $x = 4$ then $4^{-1} = \frac{1}{4}$.

Definition
If x is any number then
$x^0 = 1$.
Example
$4^0 = 1$
$8^0 = 1$
$-143.7^0 = 1$.

It is also useful, as well as knowing these definitions, to know the following rules for handling expressions with exponents in them.

Rule 1
$x^a \cdot x^b = x^{a+b}$ for any number x and any numbers a and b, i.e. when multiplying the numbers we *add* the exponents.
Example
$x^4 \cdot x^3 = x^7$.
Example
$x^3 \cdot x^{-1} = x^2$.
N. B.
1. The x number *must be the same* in both parts of the expression.
2. Although we can multiply different powers of x, we cannot add different powers.

So, for example, an expression $x^3 \cdot x^2$ can be simplified to $x^{3+2} = x^5$, but an expression like $x^3 + x^2$ cannot be simplified.

Rule 2
$$\frac{x^a}{x^b} = x^{a-b},$$
i.e. when dividing numbers we subtract exponents.
Example
$$\frac{x^3}{x^2} = x^{3-2} = x^1 = x.$$
Example
$$\frac{x^2}{x^3} = x^{2-3} = x^{-1} = \frac{1}{x}.$$

Rule 3
$$(x^a)^b = x^{a \cdot b},$$
i.e. when taking exponents of expressions already having an exponent, we *multiply* the exponents.
Example
$(x^3)^2 = x^6.$
Example
$(x^3)^{-1} = x^{-3}.$

Rule 4
$x^a \cdot y^a = (x \cdot y)^a$
N. B. Here although x and y can be different, the exponent must be the same in both parts. (Contrast this with rule 1.)
Example
$x^3 \cdot y^3 = (xy)^3.$
Example
$\sqrt{xy} = (xy)^{1/2} = x^{1/2} \cdot y^{1/2} = \sqrt{x} \cdot \sqrt{y}.$

2.9.2 Exponential functions

An expression like 2^x defines a mapping, just as much as x^2 does. (Although we might have to be a bit more careful over the choice of domain.)
If $f(x) = 2^x$ then, for example,
  $f : 1 \to 2$
  $f : 2 \to 4$
  $f : 3 \to 8$
  $f : -1 \to \frac{1}{2}$
whereas if $f(x) = x^2$

then $f: 1 \to 1$
$f: 2 \to 4$
$f: 3 \to 9$
$f: -1 \to 1$.

Definition
A function of the type $f(x) = a^x$ for a number a (where a is any number greater than 1) is called an *exponential function*.

Example
$f(x) = 5^x$ is an exponential function and
  $f(1) = 5$
  $f(2) = 5^2 = 25$
  $f(3) = 5^3 = 125$
  $f(-1) = 5^{-1} = \frac{1}{5}$
  $f(\frac{1}{2}) = 5^{1/2} = \sqrt{5} = 2.236$
  $f(0) = 5^0 = 1$.
The graph of $f(x) = 5^x$ is shown in Fig. 2.18.

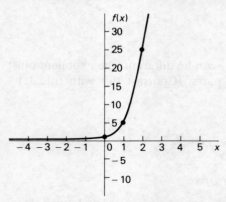

Fig. 2.18

In fact, all exponential functions have a similar graph.

N. B. Exponential functions $f(x) = a^x$ must not be confused with functions of the type $f(x) = x^a$, which are called *power* functions.

One particular exponential function is used more frequently than any other, even though it corresponds to what seems a strange choice for a. The number used is denoted by the letter 'e' and $e \simeq 2.71828$. The function $f(x) = e^x$ is sometimes referred to as the *natural* exponential function, although we shall often refer to it simply as *the* exponential function.

2.9.3 The natural exponential function

$f(x) = e^x$
or $f(x) \simeq 2.71828^x$.
So, for example,
$f(0) \simeq 2.71828^0 = 1$
$f(1) \simeq 2.71828^1 = 2.71828$
$f(2) \simeq 2.71828^2 = 7.38905$
$f(5) \simeq 2.71828^5 = 148.413$

$f(-1) \simeq 2.71828^{-1} = \dfrac{1}{2.71828} = 0.36788.$

N. B. Notice how as x increases the exponential function increases at an ever increasing rate – this kind of growth is, of course, known as *exponential growth*.

The graph of the exponential function is shown in Fig. 2.19.

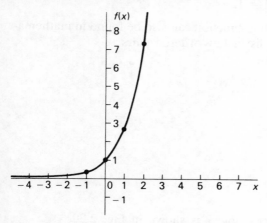

Fig. 2.19

Because the number 'e' is not a simple number, the calculation of $f(x)$ given x, is not easy. For this reason the images are often given in a set of mathematical tables, under the exponential function. We shall see below examples of other functions with ordered pairs which are included in mathematical tables.

2.9.4 The natural logarithm function

The exponential function $f(x) = e^x$ (and indeed every exponential function) is a 'one to one' function and does, therefore, have an inverse function.

This inverse function is called the *natural logarithm function* (sometimes referred to as the Naperian logarithm or hyperbolic logarithm) and we write it as $g(x) = \log_e x$, which we read as 'log of x to base e'. (The e part refers to the fact that it is the inverse function of $f(x) = e^x$. There are an infinite number of logarithm functions, e.g. $g(x) = \log_{10} x$, the so-called common logarithm function, which is used in arithmetic and which is the inverse function of $f(x) = 10^x$.)

Since $g(x) = \log_e x$ is the inverse function of the exponential function we can get its ordered pairs from the exponential function.

Example

If $f(x) = e^x$ then
$f(1) = 2.71828$
and so if $g(x) = \log_e x$
$g(2.71828) = 1$.

Ordered pairs for the $\log_e$ function can also be found in mathematical tables. Here we list a few ordered pairs.

x	$\log_e x$
1	0
2	0.6931
4	1.3863
5	1.6094
10	2.3026
0.5	−0.6931

The graph of the $g(x) = \log_e x$ is shown in Fig. 2.20.

N. B. The $\log_e x$ function is *not* defined for negative values of x or even when $x = 0$. That is, the domain of the function is the set of positive reals only.

However, it is possible to have an x such that $\log_e x$ is negative. (For example, $\log_e 0.5 = -0.6931$.)

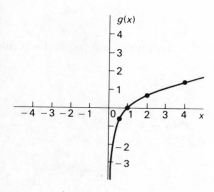

Fig. 2.20

2.9.5 Properties of logarithm functions

We have defined the natural logarithm function as being the inverse function of the natural exponential function (and other logarithm functions as being the inverse of corresponding exponential functions), but logarithm functions are important in their own right because of certain unique and important properties they possess. We shall illustrate these properties with reference to the natural logarithm function but they do hold for *any* logarithm function (and indeed are the basis for using the $\log_{10} x$ function to assist in arithmetic operations, like multiplying two numbers together).

Property 1
Given any two positive numbers x and y then
$\log_e(x \cdot y) = \log_e x + \log_e y$.
Example
If $x = 2$ and $y = 3$ then $x \cdot y = 6$
and $\log_e(6) = 1.7918$ (using natural logarithm tables).
Now $\log_e(2) = 0.6932$
and $\log_e(3) = 1.0986$.
Therefore, $\log_e(2) + \log_e(3) = 0.6932 + 1.0986$
$= 1.7918$.

Property 2
$\mathrm{Log}_e \dfrac{x}{y} = \log_e x - \log_e y$.

Example

If $x = 5$ and $y = 4$ then $\dfrac{x}{y} = 1.25$

and $\log_e(1.25) = 0.2231$.
Now $\log_e(5) = 1.6094$
and $\log_e(4) = 1.3863$.
Therefore, $\log_e(5) - \log_e(4) = 1.6094 - 1.3863$
$= 0.2231$.

Property 3
$\log_e(x^n) = n \log_e x$.
Example
If $x = 3$ and $n = 2$ then $x^n = 3^2 = 9$
and $\log_e(9) = 2.1972$.
Now $\log_e 3 = 1.0986$.
Therefore, $2 \log_e 3 = 2(1.0986) = 2.1972$.

2.9.6 The sine function

Many readers will be familiar with the idea of the sine of an angle of a triangle. The sine function discussed here is directly connected to that idea, but it will not prove useful to consider the function in that light. Instead we shall merely describe the function in terms of its ordered pairs and particularly its graph.

The sine function is a function, written $f(x) = \sin(x)$, which has, for example, the following ordered pairs:

x	$f(x) = \sin(x)$
0	0.000
0.5	0.479
1.0	0.841
1.5	0.998
2.0	0.909
3.0	0.141
−1.0	−0.841
−2.0	−0.909

and the graph is shown in Fig. 2.21.

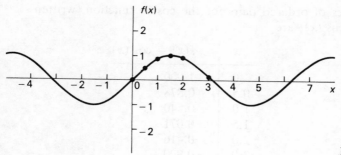

Fig. 2.21

N. B.
1. Observe how, as x increases from zero, sin (x) increases then decreases, goes negative and then starts to increase again. This 'wave-like' pattern is what makes the sine function so important.
2. Sin (x) can never be greater than 1 or less than -1 for any value of x, i.e. the image set is $\{r|r$ is a real number and $-1 \leq r \leq 1\}$.
3. The function is clearly 'many to one'. In particular there are many values of x (in fact an infinite number) which are mapped to zero. For example,
   sin $(0) = 0$
   sin $(\pi) = 0$
   where π is used to represent a number which is approximately 3.141 6
   sin $(2\pi) = 0$
   sin $(-\pi) = 0$.
4. A more complete list of ordered pairs can be found in mathematical tables.
5. Sometimes the brackets in the notation for the sine function are omitted and we write $f(x) = \sin x$.

2.9.7 *The cosine function*

A function which behaves in a very similar manner to the sine function and is often associated with it, is the cosine function. Although this can also be defined in terms of angles, it will not prove convenient to do so and we shall adopt the same procedure as above – namely, define the function by listing ordered pairs and by its graph.

Examples of ordered pairs for the cosine function, written $f(x) = \cos(x)$, are

x	$f(x) = \cos(x)$
0	1.000
0.5	0.878
1.0	0.540
1.5	0.071
2.0	−0.416
3.0	−0.990
−1.0	0.540
−2.0	−0.416

and the graph of the cosine function is shown in Fig. 2.22.

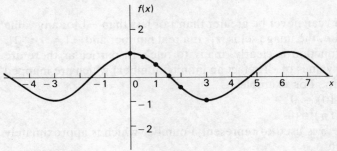

Fig. 2.22

N. B.
1. Observe again the 'wave-like' pattern, similar to that of the sine function. The only difference between the two functions is that the whole cosine function has been 'moved back' an amount equal to $\frac{1}{2}\pi \simeq 1.57$.
2. The image set is again $\{r \mid r \text{ is a real number and } -1 \leq r \leq 1\}$.
3. The cosine function is 'many to one' and in particular
   $\cos(\frac{1}{2}\pi) = 0$
   $\cos(\frac{3}{2}\pi) = 0$
   $\cos(-\frac{1}{2}\pi) = 0$.
4. A more complete list of ordered pairs can be found in mathematical tables.
5. The cosine function is often written without brackets as $f(x) = \cos x$.
6. $\cos(0) = 1$ whereas $\sin(0) = 0$.

2.10 Applications to economics

Mappings, and in particular functions, occur frequently in economic analysis in, for example, cost functions, profit functions, demand and supply functions. In this section we shall introduce functions which are used in later chapters to illustrate other mathematical techniques.

2.10.1 Demand functions

Consider the market for a single commodity. It is usually accepted that as the price of the commodity increases, people buy less of that commodity, and conversely if the price were to fall, people would buy more. We might then observe, over a period of time, the following quantities demanded by consumers at the different prices.

Price	Quantity demanded
£2	44 units
£5	35 units
£10	20 units

Notice how, as price increases, the quantity demanded decreases. Even more fundamental than this is the fact that as price *changes* so does the quantity demanded, i.e. for *different* values of price we observe *different* quantities demanded. In fact, this table of numbers is of exactly the same type as those used in section 2.5 to display the ordered pairs of mappings and functions.

If we use the letter p to denote the price, then in the spirit of section 2.5, we can write the table as

p	$f(p)$
2	44
5	35
10	20

Any value of p will have an associated value $f(p)$ because, at any level of price, there will be a certain quantity demanded. What we have is a mapping or, more specifically a function, since at each level of price there will be just one value for the quantity demanded. Such a function relating the quantity demanded for a commodity to the price of the commodity is called a *demand function*.

In many cases all we know about a demand function is limited to a few ordered pairs, as in the example above, but in some cases we may be able to determine a formula for the function.

Example
The quantity demanded of a particular commodity is related to the price (p) of that commodity by the following formula:
quantity demanded = $50 - 3p$.

Given a formula for a demand function we can calculate the quantity demanded at any level of price, by substituting the appropriate value of p into the formula.

Notation
We shall denote the quantity demanded by q^d. Hence, we can write, in the example above,
$q^d = 50 - 3p$.

Remark
The demand function given in the example above is a *linear* function because the function is of the form given in section 2.7.1. Consequently, the graph of this demand function will be a straight line (Fig. 2.23).

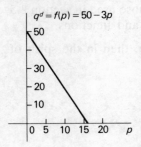

2.23

Remark
The graph of a demand function is usually expected to slope *downwards*, since as the price increases we expect the quantity demanded to decrease.

Remark

The domain of the demand functions is restricted to values of $p \geq 0$. (In fact, the domain may need to be restricted even more to prevent $f(p)$ becoming negative, since we could not have a negative quantity demanded.)

Remark

In the real world there are many factors other than price which determine the quantity of a good demanded – e.g. people's incomes, tastes, and the prices of other goods are all important. Here we are assuming that all of these remain constant for simplicity.

2.10.2 Exercises

For the following demand functions, find the quantity demanded if the price is (a) £2; (b) £4; (c) £5; (d) £10.
Hence draw the graphs of the demand functions.

(i) $q^d = 50 - p$,

(ii) $q^d = 20 - \frac{1}{2}p$,

(iii) $q^d = \dfrac{40}{p}$.

2.10.3 More on demand functions

We have considered the demand function to give the quantity which will be demanded at any particular price, i.e. given a value for p we can find the corresponding q^d. However, it is often more convenient to reverse the process, i.e. we may be given a value for q^d and asked to find what value of p will ensure that this quantity is demanded.

Example

The demand function $q^d = 40 - 2p$ can be used to find the value of q^d for any given value of p.

Alternatively, the function can be re-written as $p = 20 - \frac{1}{2}q^d$, and, in this form, can be used to find the value of p which will ensure that a given value of q^d prevails. If, for example, we wished to know what price is needed to ensure that 10 units of a commodity will be demanded, we can use this form of the demand function, substitute $q^d = 10$ and find that the price is $p = 20 - \frac{1}{2} \cdot 10 = 20 - 5 = 15$, i.e. if the price $p = 15$ then 10 units will be demanded. We can check that this is indeed the case by substituting $p = 15$ into the original form of the demand function.

Remark

The original form of the demand function, which has $q^d = f(p)$, can easily be transformed into this new form, $p = g(q^d)$, using the rules for manipulating equations, which are given in section 4.2.2.

Remark

It will be convenient to refer to the original form as the 'quantity' form of the demand function, and the new form as the 'price' form of the demand function.

Example

The 'quantity' form of a demand equation is $q^d = 100 - 4p$. The 'price' form is $p = 25 - \frac{1}{4}q^d$.

2.10.4 Supply functions

We again consider the market for a single commodity but this time examine the supply side of that market. As the price that suppliers can obtain for a commodity increases, we would expect that suppliers will be prepared to increase their output. We might observe the following quantities supplied, corresponding to the different prices.

Price	Quantity supplied
£2	7 units
£5	13 units
£10	23 units

Again we have a table identical to those used in section 2.5 to display the ordered pairs of mappings and functions, and using the letter p to denote price we could write the table as

p	$f(p)$
2	7
5	13
10	23

Any value of price will have an associated quantity supplied and therefore we have a mapping which again turns out to be a function.

Not surprisingly, this function is called a *supply function*, since it relates the quantity of the commodity supplied to the price of the commodity.

In some cases it may be possible to obtain a formula for the supply function as in the following example:

Example

The quantity of a particular commodity supplied is related to the price (p) of that commodity by the formula

Quantity supplied = $3 + 2p$.

From this formula we could obtain the quantity which would be supplied for *any* price by substitution.

Notation

We shall denote the quantity supplied by q^s. Hence we can write the formula in the example above as

$q^s = 3 + 2p$.

Remark

The supply function given in the example is a *linear* function because it is of the form given in section 2.7.1. Consequently the graph of this supply function will be a straight line (Fig. 2.24).

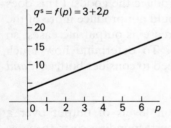

Fig. 2.24

Remark

The graph of a supply function is usually expected to slope *upwards* since as the price increases we expect the quantity supplied to increase.

Remark

The domain of supply functions is restricted to values of $p \geq 0$, and may even be more restricted to prevent q^s being negative.

Remark
As with the demand function introduced in section 2.10.1, in practice the quantity supplied will depend on factors other than price.

2.10.5 Exercises

For the following supply functions, find the quantity supplied if the price is:
   (a) £2   (b) £4   (c) £5   (d) £10.
Hence draw the graphs of these supply functions.
(i) $q^s = 1 + 3p$
(ii) $q^s = 2 + \frac{1}{2}p$
(iii) $q^s = \log_e p$
with the additional restriction that $p \geq 1$ (refer to the table in section 2.9.4).

2.10.6 Cost functions

Consider a firm which produces a single commodity. In general, the more of the commodity the firm produces, the more it will cost the firm – it will, for example, need more raw materials which must be paid for, and more men to produce the goods. (This does not mean, of course, that the firm should not produce more of the good, because although the costs increase as output increases, so does the revenue the firm receives. To determine how much output the firm should produce, we need to consider both cost *and* revenue.)

Example
A firm has produced at five different levels of output over a number of weeks and observed the corresponding costs to be as follows:

Output	Cost
4 units	£132
5 units	£145
8 units	£244
10 units	£420
20 units	£4 420

This table of numbers is identical to those introduced in section 2.5 for the ordered pairs of mappings.

If we use the letter x to denote the amount of output produced, then, in the spirit of section 2.5, we can write the table as

x	$f(x)$
4	132
5	145
8	244
10	420
20	4 420

Clearly *any* level of output will have an associated cost of producing it, i.e. for *any* value of x there will be a value of $f(x)$. Therefore we do have a mapping, or more specifically a function, since we presume that for each output level there will only be one associated cost level. (While it is conceivable that there could be more than one way of producing a certain output and hence more than one cost associated with that output, we presume that the firm will only consider the method which is associated with the lowest cost.)

The function we have introduced relating cost of producing a certain level of output, to that output, is called the *cost function* for the firm, or sometimes the *total cost function*.

In most practical cases all that the firm will know about its cost function is a few ordered pairs, as in the example above, but in some cases the firm may have a formula for the function.

Example

A firm has a total cost function given by

  cost $= x^3 - 12x^2 + 60x + 20$

where x is the number of goods produced.

In this case the firm can calculate the cost associated with *any* level of output, by substituting into the formula.

N. B. The cost function in this example is a third degree polynomial, sometimes called a 'cubic' function (see section 2.7.4).

Remark

The *domain* of cost functions will usually be restricted to the set of real numbers greater than or equal to zero (or in some cases restricted to integers, greater than or equal to zero).

From the total cost function we can define the following.
1. *The fixed cost*: The cost incurred by the firm irrespective of how much is produced. The firm will pay this cost whether it produces 100 units or 0 units. (Given the formula for a cost function, as in the above example, the easiest way to find the fixed cost, is to find the cost when output (x) is zero. In the example above, when $x = 0$, cost = 20 and therefore the fixed cost = 20.)
2. *The variable cost*: That part of the total cost which changes as output changes (as opposed to the fixed cost which does not). In fact variable cost = total cost − fixed cost. In the example above:
variable cost = $(x^3 - 12x^2 + 60x + 20) - 20$
$= x^3 - 12x^2 + 60x$.
3. *The average cost*: The cost/unit of output. If the total cost of producing 8 units is £244, then the average cost is £244/8 = £30.5 per unit. If the total cost of producing 5 units is £145, then the average cost is £145/5 = £29 per unit.

N. B. The average cost can change as output changes. That is, the average cost is also a function, in the same way as total cost is. This can be clearly seen if the total cost is given by a formula.

Example
If cost = $x^3 - 12x^2 + 60x + 20$

then average cost = cost/unit of output = $\dfrac{\text{cost}}{\text{output}}$

$$= \dfrac{x^3 - 12x^2 + 60x + 20}{x}$$

Average cost $= x^2 - 12x + 60 + \dfrac{20}{x}$

N. B. We have to divide *every* term in the numerator by x.
So the average cost is also given by a formula and we can use the formula to find the average cost at different levels of output.

2.10.7 Exercises

Given the following total cost function for a firm, find the fixed cost, and the variable and average cost formulae.
Cost = $x^3 - 10x^2 + 150x + 200$.
What are the total, fixed, variable and average costs for the following outputs?
$x = 1, 5, 10$.

2.10.8 Revenue functions

In section 2.10.6 we introduced the idea of a cost function for a firm producing a single commodity. Of course, the firm would not produce at all if there were not some benefit gained from producing, to at least offset this cost. This benefit is in the form of the revenue the firm receives when it sells its product. If we again use the letter x to denote the amount the firm produces and then sells, the revenue the firm receives will be $p \cdot x$, i.e. the number of units sold multiplied by the revenue received for each unit (the price p). The price the firm receives for each unit it sells depends on the demand function for the firm's product. Recall (from section 2.10.1) that the demand function relates the quantity of the firm's product demanded, to the price. If the firm is to sell x units of its product, there must be a demand for x units and hence the price must be that which will ensure that x units are demanded. In this case it is the 'price' form of the demand equation which is required (see section 2.10.3)

Example

A firm is faced with the following demand function for its own product $p = 100 - 4x$.

N. B. We have used x here rather than the traditional q^d associated with demand functions.

The firm's revenue from selling x units will then be
revenue = $p \cdot x$
= $(100 - 4x) \cdot x$
revenue = $100x - 4x^2$.

.This formula gives the revenue function (or total revenue function as it is sometimes called) for the firm. It is a function because for every level of output (value of x) we have a single value of revenue associated with it, which can be obtained from the formula.

The following table gives some of the possible outputs and associated revenues for the above example.

Output x	Revenue
2	184
5	400
10	600
20	400

Remark
Unlike the firm's cost function, there is no reason why the revenue function should always increase. The revenue is given by $p \cdot x$ and as output increases we might expect the price to fall if this extra output is to be sold. Consequently, although x is increasing, p is decreasing and may decrease sufficiently to offset the increase in x.

Remark
If the firm's output is only a very small part of the total market supply then any increase in its output is unlikely to affect the market price. In this case (the case of 'perfect competition') the *firm's* demand function is $p = $ a constant, (i.e. the price is *not* affected by the firm's output x) and total revenue for the firm is given by

revenue = (a constant) $\cdot x$.

In this case the firm will always increase revenue as it increases output.

Remark
We can define the *average revenue*, in the same way as we defined *average cost*, as the revenue/unit of output. This of course is exactly what is meant by the price of the product.

2.10.9 Exercises

A firm is faced with a demand function for its own product of the form $p = 40 - 2x$.

Find the total revenue function for the firm and hence the revenue the firm receives if it produces:

(a) 0 units; (b) 2 units; (c) 10 units; (d) 20 units.

2.10.10 Profit functions

We have seen in sections 2.10.6 and 2.10.8 how, as the firm varies its output, its cost and revenue also vary. In order to determine what level of output to produce the firm examines its *profit function* which gives the excess of revenue over cost, i.e. profit = revenue − cost.

We have seen how the firm's revenue and cost can be considered to be functions of output. Consequently it is of no surprise that the firm's profit can also be considered to be a function of output.

Example

A firm has a cost function given by cost = $x^3 - 12x^2 + 60x + 20$ and a revenue function, revenue = $100x - 4x^2$.

The firm's profit function is given by
$$\text{profit} = \text{revenue} - \text{cost}$$
$$= (100x - 4x^2) - (x^3 - 12x^2 + 60x + 20)$$
$$= 100x - 4x^2 - x^3 + 12x^2 - 60x - 20,$$
i.e. profit = $40x + 8x^2 - x^3 - 20$

and given a level of output (i.e. a value for x) we can calculate the firm's profit if it were to produce that level of output.

The table below gives some examples.

Output x	Profit
0	−20
2	84
5	255
10	180

We observe that there is no reason why the profit function should increase as output increases. Costs will increase and revenue may not increase sufficiently to offset this. (Indeed as we saw in section 2.10.8, revenue may actually decrease.) What the firm will probably do is to select that level of output which will make its profits as large as possible. We shall examine how the firm determines what level of output will maximise profits in section 6.6.1.

Remark

The firm may, of course, choose not to maximise its profit, but pursue some other objective. The reader is referred to, for example, Pickering's *Industrial Structure and Market Conduct* for a discussion of the different possible objectives of firms.

2.10.11 Exercise

A firm has the following cost and revenue functions:
cost $= x^3 - 10x^2 + 150x + 200$
revenue $= 210x - x^2$
where x is the number of units of output.

Find the profit function and hence the profits the firm would make if it produced
(a) 0 units, (b) 2 units (c) 10 units (d) 20 units.

Chapter 3

Operations on mappings

3.1 Introduction

We have already observed, in Chapter 1, how, given two sets, we can consider combining them to form a third set (by union or intersection) in much the same way as we can combine numbers to form a third (by addition, multiplication, etc.). Mathematicians find it natural to try and extend these ideas to every branch of the subject and, in particular, they have defined ways of combining two functions to form other functions.

For the definitions which follow we shall assume that we have two mappings, f and g, both of them functions and both mapping the set of real numbers to the set of real numbers.

3.2 Operations

We have, then, $f: R \to R$
and $\qquad\quad g: R \to R$

3.2.1 Addition of functions

Given f and g we define $f + g$ to be a function such that, if $x \in R$, then $f + g$ maps x to $f(x) + g(x)$,
i.e. $f + g: x \to f(x) + g(x)$ $(x \in R)$.
Example
If $f(x) = 2x + 1$
and $\quad g(x) = 1 - x$
then $\quad f: 2 \to 5 \quad$ or $\quad f(2) = 5$
and $\quad g: 2 \to -1 \quad$ or $\quad g(2) = -1$.

Now $\quad f + g: x \to f(x) + g(x)$
so $\quad\ \ f + g: 2 \to 5 + (-1)$
$\qquad f + g: 2 \to 4.$

Similarly, since $f: 4 \to 9$ or $f(4) = 9$
and $g: 4 \to -3$ or $g(4) = -3$
then $f + g: 4 \to 6$

The formula for the function $f + g$ is easily derived from
$f(x) = 2x + 1$
$g(x) = 1 - x$.
Since $f + g: x \to f(x) + g(x)$
then $f + g: x \to (2x + 1) + (1 - x)$
or $f + g: x \to x + 2$
or we can write it as $f + g(x) = x + 2$.

Example
If $f(x) = x^2$
and $g(x) = 3x - 1$
then $f + g: x \to x^2 + 3x - 1$
or $f + g(x) = x^2 + 3x - 1$
and so, for example, $f + g: 2 \to 2^2 + 3 \cdot 2 - 1 = 4 + 6 - 1$,
i.e. $f + g: 2 \to 9$ or $f + g(2) = 9$

In a similar fashion we can define subtraction, multiplication and division of functions.

3.2.2 Subtraction of functions

$f - g$ is a function such that if $x \in R$, then $f - g$ maps x to $f(x) - g(x)$,
i.e. $f - g: x \to f(x) - g(x)$ $(x \in R)$.

3.2.3 Multiplication of functions

$f \times g$ is a function such that if $x \in R$, then $f \times g$ maps x to $f(x) \times g(x)$,
i.e. $f \times g: x \to f(x) \times g(x)$.

3.2.4 Division of functions

$f \div g$ is a function such that if $x \in R$, then $f \div g$ maps x to
$f(x) \div g(x)$ or $\dfrac{f(x)}{g(x)}$,
i.e. $f \div g: x \to f(x) \div g(x)$.

N. B. The definition for the division of functions will break down if $g(x) = 0$. So if for some x, $g(x)$ is equal to zero, $f \div g$ will not be defined for that particular x.

Example
If $\quad f(x) = x^2 + 1$
and $\quad g(x) = 4x + 2$

then $\quad f + g : x \to (x^2 + 1) + (4x + 2)$,
i.e. $\quad f + g : x \to x^2 + 4x + 3 \quad$ or $\quad f + g(x) = x^2 + 4x + 3$

$\quad f - g : x \to (x^2 + 1) - (4x + 2)$,
i.e. $\quad f - g : x \to x^2 - 4x - 1 \quad$ or $\quad f - g(x) = x^2 - 4x - 1$

$\quad f \times g : x \to (x^2 + 1) \cdot (4x + 2)$,
i.e. $\quad f \times g : x \to 4x^3 + 2x^2 + 4x + 2 \quad$ or $\quad f \times g(x)$
$$= 4x^3 + 2x^2 + 4x + 2$$

$$f \div g : x \to \frac{(x^2 + 1)}{(4x + 2)} \quad \text{or} \quad f \div g(x) = \frac{x^2 + 1}{4x + 2}$$

Considering a particular element (x) of the domain, e.g. $x = 2$, we have $\quad f(2) = 5$
$\quad\quad\quad\quad\quad g(2) = 10$

and using the formulae derived about for each of the new functions we have
$f + g(2) = 2^2 + 4 \cdot 2 + 3 = 4 + 8 + 3 = 15$
$f - g(2) = 2^2 - 4 \cdot 2 - 1 = 4 - 8 - 1 = -5$
$f \times g(2) = 4 \cdot 2^3 + 2 \cdot 2^2 + 4 \cdot 2 + 2 = 32 + 8 + 8 + 2 = 50$
$$f \div g(2) = \frac{2^2 + 1}{4 \cdot 2 + 2} = \frac{4 + 1}{8 + 2} = \frac{5}{10} = \frac{1}{2}.$$

N. B. The function $f - g$ will not be the same as the function $g - f$. Similarly, $f \div g$ differs from $g \div f$.

Remark
We can quite naturally extend our definitions, given three functions f, g and h, to construct new functions, like, for example,
$f + g + h$
or $f + g - h$
or $(f + g) \times h$.

3.2.5 Exercises

Given three functions
$$f(x) = 3x + 2$$
$$g(x) = x^2$$
$$h(x) = 1 - x$$
what are the formulae which give the following functions?
(a) $f + g$; (b) $f \times h$; (c) $g - h$; (d) $f + g + h$;
(e) $f \times (g + h)$; (f) $(f \times g) + h$; (g) $(f - h) \div g$.

Use the formulae to find out where the number 1 is mapped to under each of the new functions.

3.3 Composition of functions (function of a function)

We have seen how we can combine two functions to produce a third, using the operations $+, -, \times, \div$, in a way which is familiar, in that it is based on the arithmetic of real numbers.

A fifth way of combining two functions is very different, this is the *composition of functions*.

Consider two functions
$$f(x) = 3x + 2$$
and $g(x) = 1 - x$.
Select a value for x, say $x = 2$.
Now $f : 2 \to 8$ or $f(2) = 8$
and $g : 2 \to -1$ or $g(2) = -1$.

It is on the basis of $f(2)$ and $g(2)$ that we have defined addition, subtraction, multiplication and division of functions. For composition of functions this is not so.

We start off, as before, with $x = 2$. Again we find the image of 2 under the f function, i.e. $f(2) = 8$, but instead of returning to 2 to find its image under g, we find the image of 8 (i.e. $f(x)$) under g, which is -7.

Symbolically $2 \xrightarrow{f} 8 \xrightarrow{g} -7$.

Generally, in taking the composition of functions we apply g to $f(x)$ not to x, i.e. $x \xrightarrow{f} f(x) \xrightarrow{g} g(f(x))$ and for any x we will have an image $g(f(x))$.

For example, if $x = 3$

$3 \xrightarrow{f} 11 \xrightarrow{g} -10$.

So the image of 3 under the combined effect of f and g is -10.

Since for every x in the domain we have an image, clearly we have defined a mapping (in fact a function). This new function is called the composition of f and g and written gof. (An easy way to remember this notation is to read it as g of f.)

Example
If $\quad f(x) = 2 + x^2$
and $\quad g(x) = x + 1$
what are the images of 2, 1, -1, under the function gof?

Consider $x = 2$ first

$2 \xrightarrow{f} 6 \xrightarrow{g} 7,$

i.e. $\quad gof: 2 \to 7$.

Now $\quad x = 1$

$1 \xrightarrow{f} 3 \xrightarrow{g} 4,$

i.e. $\quad gof: 1 \to 4$

and finally when $x = -1$

$-1 \xrightarrow{f} 3 \xrightarrow{g} 4$.

So $\quad gof: -1 \to 4$.

Let us now have a formal definition of composition.

Definition
If $f: R \to R$ and $g: R \to R$
then gof is a function such that if $x \in R$, then
$\quad gof$ maps x to $g(f(x))$,
$\quad$ i.e. $gof: x \to g(f(x))$.

Remark
The composition of functions is often referred to as the *function of a function*, because we are interested in $g(f(x))$, (i.e. a function g, not of x, but of the image of another function f).

Remark
The composition will only be defined if the domain of g is the same as the image set of f.

Remark
The function *fog* will be obtained in the reverse order, i.e. we first apply g and then f
$$x \xrightarrow{g} g(x) \xrightarrow{f} f(g(x)).$$
Example
If $f(x) = x^2$
$g(x) = 4x$
what are the images of 2 under the functions *gof* and *fog*?

Consider first *gof*
$$x \xrightarrow{f} f(x) \xrightarrow{g} g(f(x))$$
so $2 \xrightarrow{f} 4 \xrightarrow{g} 16$,

i.e. $gof: 2 \to 16$.

Now for *fog*
$$x \xrightarrow{g} g(x) \xrightarrow{f} f(g(x))$$
so $2 \xrightarrow{g} 8 \xrightarrow{f} 64$,

i.e. $fog: 2 \to 64$.
Notice that *fog* and *gof* do not produce the same answer.

3.3.1 Exercises

If $f(x) = x + 2$
and $g(x) = 3x$
what are the images of 0, 1 and 2 under the functions
$f + g$, $f \times g$, $g - f$, *gof*, *fog*?

3.4 Derivation of formula for *gof*

Consider two functions $f(x) = 3x + 2$
and $g(x) = 1 - x$.
We have defined *gof* and, given a particular element in the domain, we can find its image under the function *gof*.
Example
$2 \xrightarrow{f} 8 \xrightarrow{g} -7$,
i.e. $gof: 2 \to -7$.

What we now wish to do is to find a *formula* that will demonstrate what happens to *any* element x from the domain.

In the above example the formula turns out to be

$gof: x \rightarrow -1 - 3x$

and we can see that this does work when $x = 2$, for the formula predicts that

$gof: 2 \rightarrow -1 - 3 \cdot 2,$

i.e. $gof: 2 \rightarrow -7.$

What we shall now do is attempt to find this formula for ourselves.

$f(x) = 3x + 2$
$g(x) = 1 - x$

we want $gof: x \rightarrow ?$

In fact the image of x, under gof is $g(f(x))$, so we really want $g(f(x))$.

Now $f(x) = 3x + 2$

so $g(f(x)) = g(3x + 2).$

We know that $g(x) = 1 - x$ but we do not want $g(x)$, we want $g(3x + 2)$.

Recall that when we write $g(x) = 1 - x$ what we mean is g(any number) $= 1 -$ (that number); we used x because it was more convenient than writing 'any number', but here it will be more convenient to use 'any number'.

So g(any number) $= 1 -$ (that number).

Therefore $g(3x + 2) = 1 - (3x + 2)$
$= 1 - 3x - 2$
$= -1 - 3x.$

So we have demonstrated that

$g(f(x)) = -1 - 3x$

and consequently that

$gof: x \rightarrow -1 - 3x.$

Example

$f(x) = x - 1$
$g(x) = x^2$

Find the formula for *gof*.

We again require $g(f(x))$ and we know $f(x) = x - 1$.
Therefore we require $g(x - 1)$.
Now $g(x) = x^2$
or g(any number) $=$ (that number)^2.

Therefore $g(x - 1) = (x - 1)^2$,
i.e. $g(f(x)) = (x - 1)^2$
and $gof: x \to (x - 1)^2$.

Example
$f(x) = 3x^2$
$g(x) = 2 + 5x$
Find the formula for *gof*.

We again require $g(f(x))$
and we know $f(x) = 3x^2$
Therefore we require $g(3x^2)$.
Now $g(x) = 2 + 5x$
or g(any number) $= 2 + 5 \cdot$ (that number).
Therefore $g(3x^2) = 2 + 5 \cdot (3x^2)$
$= 2 + 15x^2$
i.e. $g(f(x)) = 2 + 15x^2$
and $gof: x \to 2 + 15x^2$.

3.4.1 Exercises

Find the formulae for *gof* for the following examples.
(a) $f(x) = 1 + 4x$
$g(x) = 3x + 2$
(b) $f(x) = 8x$
$g(x) = x^2$
(c) $f(x) = 5x^2$
$g(x) = 2x + 3$.

3.5 Derivation of formula for *fog*

Given $f(x) = 3x + 2$
and $g(x) = 1 - x$
we have shown that $gof: x \to -1 - 3x$.
We shall now derive the formula for *fog* in a similar way.

The difference between *gof* and *fog* is that in the latter case we first apply g to x and then f to $g(x)$, i.e. we require $f(g(x))$.
(Recall that for *gof* we required $g(f(x))$.)

So we require $f(g(x))$
and we know that $g(x) = 1 - x$.
Therefore we require $f(1 - x)$.
Now $f(x) = 3x + 2$
or $f(\text{any number}) = 3 \text{ (that number)} + 2$.
Therefore $f(1 - x) = 3(1 - x) + 2$
$= 3 - 3x + 2$.
Therefore $f(1 - x) = 5 - 3x$,
i.e. $f(g(x)) = 5 - 3x$
and so $fog: x \to 5 - 3x$.

Example
Given $f(x) = x^2$
$g(x) = 3x - 1$
find the formulae for *fog* and *gof*.

fog
We require $f(g(x))$
and we know $g(x) = 3x - 1$.
Therefore we require $f(3x - 1)$.
Now $f(x) = x^2$
or $f(\text{any number}) = (\text{that number})^2$.
Therefore $f(3x - 1) = (3x - 1)^2$
i.e. $f(g(x)) = (3x - 1)^2$,
and so $fog: x \to (3x - 1)^2$.

gof
We require $g(f(x))$
and we know $f(x) = x^2$.
Therefore we require $g(x^2)$.
Now $g(x) = 3x - 1$
or g (any number) $= 3$ (that number) $- 1$.
Therefore $g(x^2) = 3(x^2) - 1$
$= 3x^2 - 1$
i.e. $g(f(x)) = 3x^2 - 1$
and so $gof: x \to 3x^2 - 1$.

3.5.1 Exercises

1. For the exercises in section 3.4.1, find *fog*.
2. If $g(x) = 1 - x$ and $h(x) = 3x + 2$

then the function $f(x) = \dfrac{1-x}{3x+2}$ can be written, in terms of g and h, $f(x) = \dfrac{g(x)}{h(x)}$.

Carry out a similar procedure for the following, writing each f in terms of combinations of g and h.
(a) $f(x) = (1 - x) + (3x + 2)$
(b) $f(x) = \dfrac{3x + 2}{1 - x}$
(c) $f(x) = (3x + 2) \cdot (1 - x)$
(d) $f(x) = 3(1 - x) + 2$ (use composition of functions)
(e) $f(x) = 1 - (3x + 2)$ (use composition of functions).

Chapter 4

Equations

In section 2.10 the idea of demand and supply functions was introduced. Economists are interested in how demand and supply *interact* in the market, and how the price mechanism operates to produce *equilibrium*.

In section 4.9 equilibrium in a market model is defined and the equilibrium price and quantity determined for a simple model. The effect of taxes and subsidies on these equilibrium values is then considered. The analysis is based on the methods of simultaneous equations and the mathematics in this chapter is devoted to methods for solving equations – both single equations and simultaneous equations.

4.1 Introduction

We should by now be familiar with functions, and given a function $f(x)$ and any element in the domain (i.e. any value for x) be able to find its associated element in the codomain (i.e. its image).

For example, if $f(x) = 3x^2 - 2x + 1$
then if $x = 2$ $f(x) = 3 \cdot 4 - 2 \cdot 2 + 1$
$= 9$
and if $x = 3$ $f(x) = 22$.

We are now going to change the problem, so that given a function, we shall be *given the image* of some unknown element in the domain, and asked to *find the element in the domain*.

For example, if $f(x) = 3x^2 - 2x + 1$
what element in the domain is mapped to 9? From the previous example, the element which is mapped to 9 is known to be $x = 2$. But suppose we had not already known that, how could we have found the element?

Problems of this sort, where we are given the image of an element and asked to find the element, are essentially what *solving equations* is about. More particularly, we shall usually be concerned with finding which element in the domain has an image of *zero*.

That is, given $f(x)$, find the value (or values, if the function is 'many to one') of x that are mapped to zero.

How easily this problem can be solved depends on the type of function f. We shall consider different types of functions in the next few sections.

4.2 Linear equations

In this section we consider only functions which are linear, i.e. $f(x) = ax + b$ for some numbers a and b (see section 2.7.1).

So our problem is: find the value of x (only one since linear functions are 'one to one') which is mapped to zero under a function of the $f(x) = ax + b$ type.

Or find x such that $ax + b = 0$.

Or, as we say, 'solve the equation $ax + b = 0$'.

Example

Solve the equation $2x - 3 = 0$,

(i.e. find the value of x which under the function $f(x) = 2x - 3$, is mapped to zero).

We shall consider two *methods of solution*, i.e. two different ways of finding the value x.

4.2.1 Solution by graph

Example

Solve the equation $2x - 3 = 0$.

We draw the graph of the function $f(x) = 2x - 3$ (Fig. 4.1).

x	$f(x)$
0	-3
1	-1
2	1
-1	-5

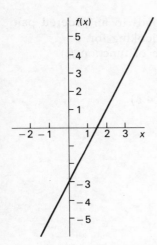

Fig. 4.1

We want the value of x which is mapped to zero, i.e. from the graph we want the point on the graph corresponding to the ordered pair $(x, 0)$ for some value of x. Such a point will, of course, be opposite the 0 on the vertical axis, and so we want the point on the graph where the graph *crosses* the horizontal axis (Fig. 4.2).

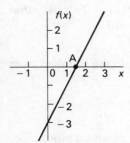

Fig. 4.2

This is the point A which corresponds to the ordered pair $(1\frac{1}{2}, 0)$. Since this point is on the graph we know that $f: 1\frac{1}{2} \to 0$, i.e. the value of x we are looking for is $x = 1\frac{1}{2}$.
(If you check by substituting $x = 1\frac{1}{2}$ into $f(x) = 2x - 3$ you *do* find that $f(1\frac{1}{2}) = 3 - 3 = 0$.)

In general then, to solve an equation $ax + b = 0$, we draw the graph of $f(x) = ax + b$ and find the point where the graph crosses

77

the horizontal axis. This point corresponds to an ordered pair $(x, 0)$ where the x is the value we are looking for.
(In fact this method works for any type of function f.)

Example

Solve $2x + 4 = 0$.

Draw the graph of $f(x) = 2x + 4$ (Fig. 4.3).

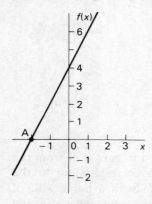

Fig. 4.3

The graph crosses the horizontal axis at the point A which corresponds to the ordered pair $(-2, 0)$.
So the solution to the equation is $x = -2$. (We can check, by substituting $x = -2$ into the function $f(x) = 2x + 4$, that $f(-2) = -4 + 4 = 0$, i.e. -2 is mapped to 0.)

Remark

This method will not be accurate unless great care is taken over drawing the graph.

4.2.2 Solution by manipulation

The following are 'allowable manipulations' of an equation, i.e. will not alter the validity of an equation:

(a) Any number can be added to (or subtracted from) *both* sides of an equation.
(b) *Both* sides of an equation can be multiplied (or divided) by any number (but see the exception below).

Remark

It must be emphasized that you must add to, subtract from, multiply or divide *both* sides of the equation.

Remark
The only *exception* is that you *cannot divide by zero*.

We shall now see how using these manipulations, we can solve our linear equations.

Example
Solve $2x - 3 = 0$.
Adding 3 to both sides $\quad 2x = 3$
Dividing both sides by 2 $\quad x = \frac{3}{2} = 1\frac{1}{2}$
which is the solution.
(The idea behind the manipulations is to isolate x on one side of the equation, leaving just a number on the other – this number then being the solution.)

Example
Solve $2x + 4 = 0$.
Subtract 4 from both sides $\quad 2x = -4$
Divide both sides by 2 $\quad x = -2$
which is the solution.

Example
Solve $4 - \frac{1}{2}x = 0$.
Add $\frac{1}{2}x$ to both sides to give $\quad 4 = \frac{1}{2}x$
Multiply both sides by 2 to give $\quad 8 = x \quad$ or $\quad x = 8$
which is the solution.

4.2.3 Exercises

1. Solve the following equations by graph and by manipulation:
   (a) $7x + 14 = 0$
   (b) $x - 4 = 0$
   (c) $12x + 2 = 0$
   (d) $4 - x = 0$
   (e) $3 - \frac{1}{3}x = 0$.
2. Using the rules for manipulating equations, show that the demand function $q^d = 100 - 4p$ can be written in its 'price form' as $p = 25 - \frac{1}{4}q^d$ (see section 2.10.3).

4.2.4 Final remark

An equation of the form $3x - 2 = 4$ can be interpreted as
either: given $f(x) = 3x - 2$ find the value of x which is mapped by f to 4

or: by simple manipulation the equation becomes $3x - 6 = 0$ and it can be thought of as: given $g(x) = 3x - 6$ find the value of x which is mapped by g to 0.

The two ways will, of course, give different graphs (in the first you draw the graph of $f(x) = 3x - 2$, in the second $g(x) = 3x - 6$) but they will give the same solution $x = 2$. Solution by manipulation is, of course, straightforward.

4.3 Quadratic equations

As the name suggests, we are interested in finding the value of x which is mapped to zero, when the function f is quadratic.
That is, $f(x) = ax^2 + bx + c$ for some a, b or c (see section 2.7.3).
Example
$4x^2 - 4x - 3 = 0$
is a quadratic equation and solving it consists of finding the value (or, in this case, values, since quadratic functions are 'many to one') of x which is mapped to zero.

We shall consider three methods of solution.

4.3.1 Solution by graph

The principle is the same as for linear equations. Draw the graph of the quadratic function and see where it crosses the horizontal axis.
Example
Solve $4x^2 - 4x - 3 = 0$.
The graph of $f(x) = 4x^2 - 4x - 3$ is shown in Fig. 4.4.

x	$f(x)$
0	-3
1	-3
2	5
3	21
-1	5

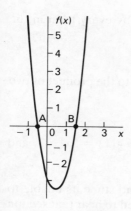

Fig. 4.4

The graph crosses the horizontal axis at *two* points A and B which correspond to the pairs $(-\frac{1}{2}, 0)$ and $(1\frac{1}{2}, 0)$, and hence we have *two* solutions $x = -\frac{1}{2}$ and $x = 1\frac{1}{2}$. That is, both $x = -\frac{1}{2}$ and $x = 1\frac{1}{2}$ are mapped to zero under f. (This is possible since f is quadratic and hence 'many to one'.)

Remark

This example illustrates the problem with accuracy in solving equations by graph. Can we be sure that the graph crosses the horizontal axis at the point $(-\frac{1}{2}, 0)$ and not $(-\frac{3}{8}, 0)$ or even $(-\frac{101}{200}, 0)$? Because of this problem we prefer the algebraic methods discussed in the next two sections to this geometric method, although solving by graph is useful *intuitively*.

4.3.2 Solving by factorisation

The general idea is to write our quadratic function $f(x)$ as a product of two linear functions.

Example

The function $f(x) = x^2 - x - 2$
can be written as $g(x) \cdot h(x)$ where $g(x) = x - 2$ and $h(x) = x + 1$.

That is $x^2 - x - 2 = (x - 2) \cdot (x + 1)$.

(The reader can easily check this is so, by 'expanding' the expression $(x - 2) \cdot (x + 1)$, and showing that it is equal to $x^2 - x - 2$. To expand an expression involving brackets, every term in the first bracket must be multiplied by every term in the second bracket.

So $(x - 2) \cdot (x + 1) = x^2 + x - 2x - 2$
$= x^2 - x - 2$.)

This process of reducing a quadratic function to the product of two linear functions is called *factorisation*.

Example

The function $f(x) = 4x^2 - 4x - 3$
can be *factorised* into $g(x) \cdot h(x)$ where $g(x) = 2x - 3$ and $h(x) = 2x + 1$,
i.e. $4x^2 - 4x - 3 = (2x - 3) \cdot (2x + 1)$

(The 'art' of factorising is not an easy one and students unable to 'spot' the *factors* $g(x)$ and $h(x)$, will be pleased to hear that section 4.3.3 gives an alternative method of solving quadratic equations.)

Having done the factorisation, the following property of numbers is used.

If a and b are numbers such that $a \cdot b = 0$, then either $a = 0$ or $b = 0$. (Because if you multiply two *non*-zero numbers together you get a non-zero number. Therefore if the product *is* zero at least one of the original numbers must be.)

Returning to our quadratic equation:
$4x^2 - 4x - 3 = 0$,
we know that $4x^2 - 4x - 3 = (2x - 3) \cdot (2x + 1)$.
Therefore we want
$(2x - 3) \cdot (2x + 1) = 0$
But using our property above, this means that
either $2x - 3 = 0$
or $2x + 1 = 0$
and we have *two* linear equations which can be solved by manipulation to give
either $x = \frac{3}{2}$
or $x = -\frac{1}{2}$, which are the two solutions.

Let us illustrate the method with another example:

Example

Solve $x^2 - x - 2 = 0$.
We can factorise as $x^2 - x - 2 = (x - 2) \cdot (x + 1)$
and so we can write $(x - 2) \cdot (x + 1) = 0$.

Therefore *either* $x - 2 = 0$
          or $x + 1 = 0$.
Again we have two linear equations which can be solved by manipulation to give
either $x = 2$
or $x = -1$, which are our two solutions to $x^2 - x - 2 = 0$.

Remark
If you can factorise the quadratic function easily the method is simple. However, some students will find factorisation more difficult than others. Those students should use the method of section 4.3.3. Students who *can* factorise must also be prepared to use the method in section 4.3.3 sometimes, since not all quadratic functions can easily be factorised.
(For example, $f(x) = x^2 + 3x + 1$
has approximate factors $g(x) = x + 0 \cdot 381\,966$ and
$$h(x) = x + 2 \cdot 618\,034$$
which will not easily be spotted!)

4.3.3 Solving by formula

Proposition
The solutions of a quadratic equation $ax^2 + bx + c = 0$ are given by the formula:

$$x = \frac{-b \pm \sqrt{b^2 - 4 \cdot a \cdot c}}{2a}$$

Remark
We shall not attempt to prove the validity of this formula. (The interested reader is referred to *An Introduction to Mathematics for Students of Economics* by J. Parry Lewis.)

Remark
The $\pm$ sign in the middle of the formula means that we can take *either* the positive value of the square root *or* the negative value. It is this which gives us our two solutions.

Remark
It is crucially important to ensure that the *a*, *b* and *c* numbers are correctly assigned, i.e. the *a* number is the number in front of x^2,

the b number the number in front of x and the c number the number on its own.

Example
Solve $2x^2 + 5x + 2 = 0$.
Here $a = 2$, $b = 5$ and $c = 2$.
The solutions are given by substituting the a, b and c values into the formula,

$$x = \frac{-5 \pm \sqrt{5^2 - 4 \cdot 2 \cdot 2}}{2 \cdot 2}$$

$$= \frac{-5 \pm \sqrt{25 - 16}}{4}$$

$$= \frac{-5 \pm \sqrt{9}}{4}$$

$$= \frac{-5 \pm 3}{4}$$

The two solutions are *either* $x = \dfrac{-5 + 3}{4} = \dfrac{-2}{4} = -\dfrac{1}{2}$

or $x = \dfrac{-5 - 3}{4} = \dfrac{-8}{4} = -2$.

Example
Solve $4x^2 - 4x - 3 = 0$.
Here $a = 4$, $b = -4$, $c = -3$.
So the solutions, using the formula are:

$$x = \frac{-(-4) \pm \sqrt{(-4)^2 - 4 \cdot 4 \cdot (-3)}}{2 \cdot 4}$$

$$= \frac{+4 \pm \sqrt{16 + 48}}{8}$$

$$= \frac{4 \pm \sqrt{64}}{8}$$

$$= \frac{4 \pm 8}{8}.$$

The two solutions are *either* $x = \dfrac{4 + 8}{8} = \dfrac{12}{8} = \dfrac{3}{2}$

or
$$x = \frac{4-8}{8} = \frac{-4}{8} = -\frac{1}{2}.$$

Example
Solve $x^2 + 3x + 1 = 0$.
Here $a = 1$, $b = 3$, $c = 1$.
So the solutions are given by:

$$x = \frac{-3 \pm \sqrt{3^2 - 4 \cdot 1 \cdot 1}}{2 \cdot 1}$$

$$= \frac{-3 \pm \sqrt{9 - 4}}{2}$$

$$= \frac{-3 \pm \sqrt{5}}{2}$$

$$\simeq \frac{-3 \pm 2 \cdot 236\,068}{2}$$

($\simeq$ means 'approximately equal to')
which gives two solutions,

either $\quad x = \dfrac{-3 + 2 \cdot 236\,068}{2} = \dfrac{-0 \cdot 763\,932}{2} = -0 \cdot 381\,966$

or $\quad x = \dfrac{-3 - 2 \cdot 236\,068}{2} = \dfrac{-5 \cdot 236\,068}{2} = -2 \cdot 618\,034$.

Example
Solve $3 + 4x + x^2 = 0$.
Here $a = 1, b = 4, c = 3$. (Remember a is the number attached to x^2, b the number attached to x and c the number left.)
So the solutions, using the formula are:

$$x = \frac{-4 \pm \sqrt{4^2 - 4 \cdot 1 \cdot 3}}{2 \cdot 1}$$

$$= \frac{-4 \pm \sqrt{16 - 12}}{2}$$

$$= \frac{-4 \pm \sqrt{4}}{2}$$

$$= \frac{-4 \pm 2}{2}.$$

The solutions are *either* $x = \dfrac{-4 + 2}{2} = \dfrac{-2}{2} = -1$

or $x = \dfrac{-4 - 2}{2} = \dfrac{-6}{2} = -3.$

4.3.4 Exercises

Solve the following equations by *all three methods*:
1. $x^2 + 3x + 2 = 0$
2. $x^2 - 2x - 3 = 0$
3. $2x^2 - 5x - 3 = 0$
4. $x^2 - 1 = 0.$

Remark
Usually quadratic equations *cannot* be solved by manipulation although in certain cases manipulation does work – see, for example, section 4.3.4 question 4.

Remark
An equation of the form $x^2 - 2x = 3$ *cannot* be solved by factorisation or formula without first getting all the terms on to one side of the equation by manipulation, i.e. re-writing as $x^2 - 2x - 3 = 0$.

4.4 Types of solution for quadratic equations

It is interesting to examine more closely the types of solution which can occur with quadratic equations. There are three different types which can occur and we shall illustrate each with an example.

4.4.1 Two unequal, real solutions

Example
Solve $6x^2 - x - 2 = 0$.
If we draw the graph of the function $f(x) = 6x^2 - x - 2$, we find

that it crosses the horizontal axis at two different places, giving us two different real solutions (Fig. 4.5).

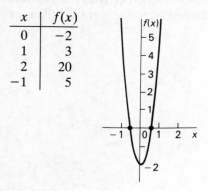

x	$f(x)$
0	-2
1	3
2	20
-1	5

Fig. 4.5

In fact the solutions are $x = -\frac{1}{2}$ or $x = \frac{2}{3}$.

If we solve the equation by formula we get:

$$x = \frac{-(-1) \pm \sqrt{(-1)^2 - 4 \cdot 6(-2)}}{2 \cdot 6}$$

$$= \frac{1 \pm \sqrt{1 + 48}}{12}$$

$$= \frac{1 \pm \sqrt{49}}{12}$$

(Notice that the $b^2 - 4ac$ term in the square root is positive.)

Therefore $x = \dfrac{1 \pm 7}{12}$

either $\quad x = \dfrac{1 + 7}{12} = \dfrac{2}{3}$

or $\quad x = \dfrac{1 - 7}{12} = -\dfrac{1}{2}.$

Whenever $b^2 - 4ac$ is greater than zero we shall get two different real numbers as solutions – one when we take the positive square root, the other when we take the negative square root.

Summary
A quadratic equation can have two unequal real solutions. This will occur when the graph of the function crosses the horizontal axis in two different places and when $b^2 - 4ac$ is greater than zero.

4.4.2 One real solution

Example
Solve $x^2 - 4x + 4 = 0$.
Solution by graph: $f(x) = x^2 - 4x + 4$

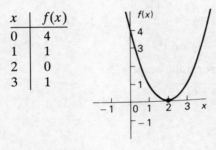

x	$f(x)$
0	4
1	1
2	0
3	1

Fig. 4.6

The graph of $f(x)$ touches the horizontal axis at *only one* point, at (2, 0) (Fig. 4.6).
Consequently there is only one solution, $x = 2$.

Solution by formula.
The solutions are:

$$x = \frac{-(-4) \pm \sqrt{(-4)^2 - 4 \cdot 1 \cdot 4}}{2 \cdot 1}$$

$$= \frac{4 \pm \sqrt{16 - 16}}{2}$$

$$= \frac{4 \pm \sqrt{0}}{2}.$$

(Notice that the $b^2 - 4ac$ term is zero.)

Therefore $x = \dfrac{4 \pm 0}{2}$

either $\quad x = \dfrac{4 + 0}{2} = 2$

or $\quad x = \dfrac{4 - 0}{2} = 2,$

i.e. we get only one solution, $x = 2$.

Summary
A quadratic equation can have just one real solution. This will occur when the graph of the function touches the horizontal axis at just *one* point and when $b^2 - 4ac$ is equal to zero.

4.4.3 No real solutions
Example
Solve $x^2 + 2x + 2 = 0$.
Solution by graph: $f(x) = x^2 + 2x + 2$

x	$f(x)$
0	2
1	5
2	10
-1	1
-2	2

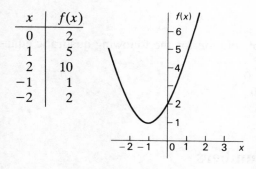

Fig. 4.7

The graph of $f(x)$ does not cross or touch the horizontal axis at any point! Therefore there is no solution (Fig. 4.7), i.e. there is no value of x which is mapped to 0 under f.

Solution by formula. The solutions are given by:

$$x = \frac{-2 \pm \sqrt{2^2 - 4 \cdot 1 \cdot 2}}{2 \cdot 1}$$

$$= \frac{-2 \pm \sqrt{4 - 8}}{2}$$

$$= \frac{-2 \pm \sqrt{-4}}{2}.$$

(Notice that $b^2 - 4ac$ is less than zero.)

We are now stuck because we cannot find the square root of a negative number:

$\sqrt{-4} \neq +2$ because $(+2) \cdot (+2) = +4$
nor is $\sqrt{-4} \neq -2$ because $(-2) \cdot (-2) = +4$.

Indeed, there is no real number, which when squared gives -4, and so $\sqrt{-4}$ is not equal to any real number.

Consequently, we cannot find a real number solution to the equation.

Summary

A quadratic equation may have no real solutions. This will occur when the graph of the function does not cross or touch the horizontal axis at any point, and when $b^2 - 4ac$ is *less* than zero.

4.4.4 Exercises

Investigate the type of solution for the following quadratic equations.

1. $3x^2 - 4x - 4 = 0$
2. $x^2 - 6x + 9 = 0$
3. $x^2 + x + 1 = 0$.

4.5 Complex numbers

In practice, square roots of negative numbers occur fairly frequently in mathematics and it is convenient to have a common, accepted way of handling them.

It is conventional, in mathematics, to denote $\sqrt{-1}$ by the letter i, and consequently, for example,

$$\sqrt{-4} = \sqrt{4 \cdot -1} = \sqrt{4} \cdot \sqrt{-1} \text{ (see rule 4 in section 2.9.1)}$$

$$= 2 \cdot i = 2i.$$

Similarly, $\sqrt{-9} = 3i$.

Numbers such as $\sqrt{-4}$ (or $2i$) are not real numbers, since no real number when squared will produce -4, they are called *imaginary numbers*.

If we return to our quadratic equation in section 4.4.3, which was 'Solve $x^2 + 2x + 2 = 0$'

our solutions were $x = \dfrac{-2 \pm \sqrt{4-8}}{2}$,

i.e. $x = \dfrac{-2 \pm \sqrt{-4}}{2}$

which we can now write as $x = \dfrac{-2 \pm 2i}{2}$

So *either* $x = \dfrac{-2 + 2i}{2} = -1 + 1i = -1 + i$

or $x = \dfrac{-2 - 2i}{2} = -1 - 1i = -1 - i.$

Numbers such as these, composed of a real part, -1 here, and an imaginary part, i here, are called *complex numbers*.

(In fact, a real number like 4 can be thought of as a complex number $4 + 0 \cdot i$, and hence the set of real numbers R, can be thought of as a subset of the set of complex numbers.)

Remark
If the solutions of a quadratic are complex they will always appear in *conjugate pairs*, i.e. if $a + bi$ is a solution (where a and b are two real numbers) then so is $a - bi$. In the example above our two solutions are $-1 + i$ and $-1 - i$ which are a conjugate pair.

Remark
The introduction of the letter *i* does not overcome the problem of the square root of a negative number, it merely leaves the square root alone as *bi*.

4.6 Higher degree polynomial equations

We wish to find the value of x which is mapped to zero by a function which is a polynomial (of degree 3 or more).
Example
Solve $x^3 + 2x^2 - 11x + 6 = 0$.

4.6.1 Solution by graph

The method used for linear and quadratic equations will again work for higher degree equations.
Example
Solve $x^3 - 4x^2 + x + 6 = 0$.
The graph of $f(x) = x^3 - 4x^2 + x + 6$ is shown in Fig. 4.8,

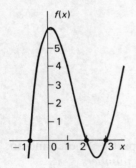

Fig. 4.8

and crosses the horizontal axis in three places to give three solutions $x = -1$, $x = 2$, or $x = 3$.

This method of solution will work for any function, but there is, of course, the problem of accuracy.

4.6.2 Solution by factorisation

Example
Solve $x^3 - 4x^2 + x + 6 = 0$.

The objective is to factorise the function $f(x) = x^3 - 4x^2 + x + 6$ into the product of two functions, a linear and a quadratic. In this example $f(x)$ will factorise into $g(x) \cdot h(x)$
where $g(x) = x - 2$ and $h(x) = x^2 - 2x - 3$,
i.e. $(x^3 - 4x^2 + x + 6) = (x - 2) \cdot (x^2 - 2x - 3)$.
Then proceeding as for the solution of quadratic equations we can argue that:
since $\quad x^3 - 4x^2 + x + 6 = 0$
then $\quad (x - 2) \cdot (x^2 - 2x - 3) = 0$.
So *either* $\quad x - 2 = 0$
$\quad$ *or* $\quad\quad x^2 - 2x - 3 = 0$
and we have reduced the problem from that of solving a third degree polynomial equation to one of solving two equations – a linear and a quadratic.
So *either* $\quad x - 2 = 0$, i.e. $\quad x = 2$
$\quad$ *or* $\quad\quad x^2 - 2x - 3 = 0 \quad$ which when solved gives $x = 3$ or $x = -1$.

Generally the objective is to factorise the high degree polynomial function into the product of polynomial functions of lower degree, and consequently solve simpler equations. As with quadratic functions, the problem with this method is that the factors may not be easy to discover.

More general methods for solving such equations are beyond the scope of this book.

4.6.3 Exercise

Solve the following equation by the graph and factorisation methods:
$\quad 8x^3 - 12x^2 - 2x + 3 = 0$.
(Hint $8x^3 - 12x^2 - 2x + 3 = (2x - 1) \cdot (4x^2 - 4x - 3)$.)

4.7 Inequalities

We have been considering in the previous section equations or equalities. That is, given a function $f(x)$, find the value, or values, of x which are mapped to some number (usually zero).

Example
'Solve $3x + 4 = 0$' means find the value of x which is mapped to zero under the function $f(x) = 3x + 4$.

We shall now look at *inequalities*, i.e. problems of the sort – 'Given a function $f(x)$ find the value, or values, of x which are mapped to numbers *less than or equal* to some number (often zero).'

Example

'Solve $3x + 4 \leq 0$' means find the values of x which are mapped to numbers less than or equal to zero, under the function $f(x) = 3x + 4$. (Note the use of the symbol '$\leq$' to denote 'less than or equal to'. The symbol '$<$' on its own represents 'less than'.)

As with equations, the ease with which such inequality problems can be solved depends on the type of function involved.

4.7.1 Inequalities involving linear functions

The methods of solution are similar to those for linear equations, i.e. solution by graph or manipulation.

Solution by graph
Example
Solve $3x + 4 \leq 0$.
The graph of $f(x) = 3x + 4$ is shown in Fig. 4.9.

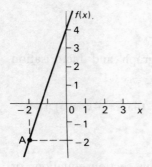

Fig. 4.9

We now require those values of x which are mapped to numbers less than or equal to zero, i.e. we are interested in that part of the graph *below* the horizontal axis, because any point on this part of the graph is opposite a negative value on the vertical axis and hence corresponds to an ordered pair which has the second element less than zero. For example, the point A is on that part of the graph below the horizontal axis and corresponds to the ordered pair $(-2, -2)$.

So we are interested in those x values which are associated with that part of the graph below the horizontal axis.

In our example it is the x values marked in a heavier line which we are interested in, i.e. the values below $-\frac{4}{3}$ (Fig. 4.10).

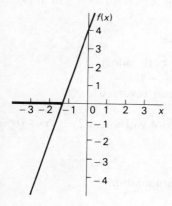

Fig. 4.10

Any value on the heavy line, e.g. $x = -3$, is associated with a point on the graph below the horizontal axis and hence with a negative element in the codomain. (In the case $x = -3$, it is mapped to -5.)

Since we are interested in any x such that $3x + 4 \leq 0$, we also include the point where the graph crosses the x axis because this is the x where $3x + 4 = 0$. This value is in fact $x = -\frac{4}{3}$.

In summary, our solution to 'solve $3x + 4 \leq 0$' is all numbers less than or equal to $-\frac{4}{3}$.

Solution by manipulation

The rules which allow us to manipulate equations (see section 4.2.2) can be used to manipulate inequalities with one very important difference. *When an inequality is multiplied or divided by a negative number, the sign of the inequality is reversed.*

Example

If $x \leq 4$ then, multiplying both sides by -1 we get $-x \geq -4$ where '$\geq$' stands for 'is greater than or equal to'.

Using manipulations we can solve inequalities in the same way as equations.

Example
Solve $3x + 4 \leq 0$.
Then $3x \leq -4$ (subtracting 4 from both sides)
and $x \leq -\frac{4}{3}$ (dividing both sides by 3)
which is the solution.
That is, *any* x which is less than or equal to $-\frac{4}{3}$ will make $3x + 4 \leq 0$.

Example
Solve $2 - \frac{1}{2}x \leq 0$.
Then $-\frac{1}{2}x \leq -2$ (subtracting 2 from both sides)
and $\frac{1}{2}x \geq 2$ (multiplying both sides by -1).
(Note how the inequality sign has been reversed).
Hence $x \geq 4$ which is the solution.
That is, *any* x greater than or equal to 4 will make $2 - \frac{1}{2}x \leq 0$.

4.7.2 Exercises

Solve the following by graph and manipulation.
1. $2x - 1 \leq 0$
2. $4 - x \leq 0$
3. $3 + 2x \geq 0$
4. $2x + 4 \leq 1$.

4.7.3 Strict inequalities

We have already introduced '<' as meaning 'is less than'. Similarly, '>' means 'is greater than'.
Strict inequalities of the form
solve $3x + 4 < 0$
or
solve $2x - 1 > 0$
can be solved in a similar way to that suggested for ordinary inequalities in section 4.7.1.

4.7.4 Non-linear inequalities

The solution of non-linear inequalities by graph is straightforward, but solution by algebraic means is considerably more difficult and we shall not attempt it here.

Example
Solve $x^2 + x - 2 \leq 0$.
The graph of $f(x) = x^2 + x - 2$ is shown in Fig. 4.11.

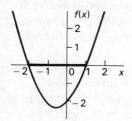

Fig. 4.11

We are interested in that part of the graph below the horizontal axis, and the values of x corresponding to it, which are shown by a heavier line, i.e. values between -2 and 1. Since $x^2 + x - 2 \leq 0$ is *not* a strict inequality, i.e. we do allow $x^2 + x - 2 = 0$, we are also interested in where the graph crosses the x axis (for there $x^2 + x - 2 = 0$), i.e. $x = -2$ and $x = 1$.
So the solution to $x^2 + x - 2 \leq 0$ is *all* values of x between -2 and $+1$, including -2 and $+1$, i.e. *all* x such that $-2 \leq x \leq 1$.

4.7.5 Exercises

Solve the following by graph:
1. $x^2 + x - 2 \leq 0$
2. $x^2 + x - 2 \geq 0$
3. $x^3 - 6x^2 + 11x - 6 \leq 0$
4. $x^2 + x + 1 \leq 0$.

4.8 Simultaneous equations

4.8.1 Introduction

In the preceding sections of this chapter we have been concerned with examining a single function $f(x)$ and attempting to find the value (or values) of x which is mapped to, for example, zero (or in the case of inequalities to numbers less than or equal to zero) under f. In this section we consider a different problem.

The first difference is that we will be concerned with *two* functions $f(x)$ and $g(x)$.

The second difference is that we are not concerned with finding the x values which are mapped to a particular element in the codomain, indeed any element in the codomain will do, providing it is the *same* element for both functions. Let us be more precise.

Definition
A *simultaneous equation* problem is of the following sort:
Given two functions $f(x)$ and $g(x)$, find the value (or values) of x which is mapped to the *same* element under both functions.

Example
If $\quad f(x) = 3x - 1$
and $\quad g(x) = -2x + 4$
$x = 2$ is *not* mapped to the same element under both functions, since
$$f : 2 \to 5$$
and $\quad g : 2 \to 0.$

But $x = 1$ *is* mapped to the same element under both functions, for
$$f : 1 \to 2$$
and $\quad g : 1 \to 2.$

So $x = 1$ is a solution of our simultaneous equation problem.

Example
Solve simultaneously $\quad f(x) = 1 + 4x$
and $\hspace{5.5em} g(x) = 11 - x$
has a solution $x = 2$ because 2 is mapped to the same element (9) under both functions.

We need a method for determining the solutions of simultaneous equations, and we shall consider two such methods.

4.8.2 Graph method

Example
Solve simultaneously
$$f(x) = 3x - 1$$
and $\quad g(x) = -2x + 4.$
We draw the graphs of $f(x)$ and $g(x)$ on the same diagram (Fig. 4.12).

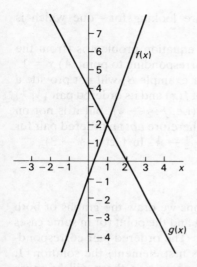

Fig. 4.12

Recall that the graph of $f(x)$ represents all the ordered pairs for the $f(x)$ mapping.

Similarly, the graph of $g(x)$ represents all the ordered pairs for the $g(x)$ mapping. We are interested in finding an ordered pair which is common to both the $f(x)$ and $g(x)$ mappings. Such a pair will clearly lie on both the $f(x)$ graph and the $g(x)$ graph, i.e. we want a point which is common to both graphs.

Such a point is the point A, where the graphs intersect (Fig. 4.13).

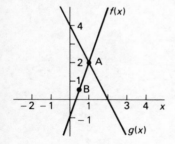

Fig. 4.13

The point A lies on the $f(x)$ graph and represents an ordered pair (1, 2) which is an ordered pair for $f(x)$ (i.e. $f : 1 \rightarrow 2$). But also A lies on the $g(x)$ graph and represents an ordered pair (1, 2) which is an ordered pair for $g(x)$ (i.e. $g : 1 \rightarrow 2$). So the point A

99

represents an ordered pair we are looking for – one which is common to both $f(x)$ and $g(x)$.

A solution to our simultaneous equation problem is (from the first element of the ordered pair corresponding to point A) $x = 1$.

N. B. Any point other than A, for example B, will not provide a solution. B is on the graph of $f(x)$ and its ordered pair $(\frac{1}{2}, \frac{1}{2})$ is an ordered pair for $f(x)$ (i.e. $f : \frac{1}{2} \to \frac{1}{2}$), but it is not on $g(x)$ and its ordered pair is therefore *not* an ordered pair for $g(x)$, (i.e. g does *not* map $\frac{1}{2} \to \frac{1}{2}$. In fact $g : \frac{1}{2} \to 3$).

Summary

To solve the simultaneous equations we draw the graphs of both functions on the same diagram and find the point (or in some cases points) where the graphs intersect. The ordered pair corresponding to this point gives us (from its first element) the solution. If there is more than one point of intersection there will be more than one solution.

Example

Solve simultaneously
$$f(x) = 2x + 1$$
and $g(x) = x^2 + x - 1$.
Draw the graphs (Fig. 4.14).

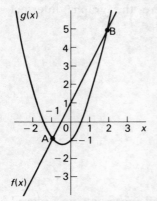

Fig. 4.14

The graphs intersect at two points A and B. The ordered pair corresponding to A is $(-1, -1)$ and the first element of this gives us a solution $x = -1$.

The ordered pair corresponding to B is (2, 5) and the first element gives us another solution $x = 2$.

So this simultaneous equation problem has two solutions $x = -1$ and $x = 2$. $x = -1$ is mapped to the same element under $f(x)$ and $g(x)$

$f : -1 \to -1$
and $g : -1 \to -1$

and $x = 2$ is also mapped to the same element under both $f(x)$ and $g(x)$

$f : 2 \to 5$
and $g : 2 \to 5$.

N. B. It is not necessary that $x = -1$ be mapped to the same element as $x = 2$.

4.8.3 Algebraic method

The idea is to replace the two functions, or the pair of simultaneous equations as they are often called, by a single equation which we have seen how to solve earlier in the chapter.

Example

Solve simultaneously
$$f(x) = 3x - 1$$
and $g(x) = -2x + 4$.

We require a value of x such that $f(x) = g(x)$, i.e. we require an x such that
$$3x - 1 = -2x + 4.$$

This is a kind of linear single equation (see section 4.2), the only difference being that we have x terms on both sides, but it can be solved by manipulation in exactly the same way as in section 4.2.2.

$$3x - 1 = -2x + 4.$$

Add $2x$ to both sides (x is just a number, as yet unknown, but still a number, so $2x$ is just a number).

Therefore $\qquad 5x - 1 = 4$
add 1 to both sides $\qquad 5x = 5$
divide both sides by 5 $\qquad x = 1$,
i.e. when $x = 1$, $3x - 1 = -2x + 4$
or $f(x) = g(x)$.

Therefore $x = 1$ is a solution.

Example
Solve simultaneously
$$f(x) = x - 1$$
and $g(x) = 2 - x$.
We require x such that $f(x) = g(x)$,
i.e. we require x such that $x - 1 = 2 - x$
which is a single linear equation.
Add x to both sides $\quad 2x - 1 = 2$
Add 1 to both sides $\quad\quad 2x = 3$
Divide both sides by 2 $\quad\; x = \frac{3}{2}$,
i.e. a solution to the simultaneous equations is $x = \frac{3}{2}$.

Example
Solve simultaneously
$$f(x) = 2x + 1$$
and $g(x) = x^2 + x - 1$.
We want x such that $f(x) = g(x)$.
Therefore we want x such that $2x + 1 = x^2 + x - 1$.

This is, in fact, a quadratic equation, which will be more easily recognisable if we do some manipulations.

Subtract $2x$ from both sides $\quad 1 = x^2 - x - 1$.
Subtract 1 from both sides $\quad\;\; 0 = x^2 - x - 2$
or, if we prefer, $\quad\quad\quad\quad\quad\; x^2 - x - 2 = 0$.

This is now clearly recognisable as a quadratic equation, and we can solve it by, for example, formula, as

$$x = \frac{-(-1) \pm \sqrt{(-1)^2 - 4 \cdot 1 \cdot (-2)}}{2 \cdot 1}$$

$$= \frac{1 \pm \sqrt{1 + 8}}{2}$$

$$= \frac{1 \pm \sqrt{9}}{2}$$

$$= \frac{1 \pm 3}{2}.$$

Either $\quad x = \dfrac{1 + 3}{2} = 2$

or $\quad\quad x = \dfrac{1 - 3}{2} = -1$.

Therefore we have two solutions to our simultaneous equation problem
$x = 2$ or $x = -1$.

N. B. When we solve the quadratic equation whether by factorisation or formula, we must first get all the terms on one side of the equation as we did above. That is, it is not possible to solve $2x + 1 = x^2 + x - 1$ directly, we must manipulate it into the $x^2 - x - 2 = 0$ form first.

4.8.4. Exercises

Solve the following simultaneous equations by the graph and algebraic methods:
1. $f(x) = x - 5$
   $g(x) = 1 - 2x$
2. $f(x) = x + 1$
   $g(x) = 2x - 1$
3. $f(x) = x^2 + 2x - 1$
   $g(x) = x + 1$
4. $f(x) = x^2 + x + 1$
   $g(x) = 2x - 2$.

4.8.5 Notation

Simultaneous equation problems appear in various disguises. A problem like
solve simultaneously $f(x) = 3x - 1$
and $g(x) = -2x + 4$
can be written as
solve $y = 3x - 1$
       $y = -2x + 4$.
(Here the $f(x)$ and $g(x)$, since we wish them to be the same number, are denoted by the same letter y.)
Or even
solve $y - 3x + 1 = 0$
       $y + 2x - 4 = 0$.
(Here the previous two equations have been manipulated, and our original f and g functions almost lost.)

Example

Solve $\begin{cases} x - y + 1 = 0 \\ 2x + y - 4 = 0 \end{cases}$

We first manipulate each equation in turn (using the rules given in section 4.2.2) until they are in the form presented previously (i.e. with y on its own on the left-hand side of each equation).

$$\begin{cases} -y = -x - 1 \\ y = 4 - 2x \end{cases}$$

or

$$\begin{cases} y = x + 1 \\ y = 4 - 2x. \end{cases}$$

We can then equate the right-hand sides of the two equations to give

$$x + 1 = 4 - 2x$$

and hence $\quad 3x = 3$

and $\quad\quad\quad x = 1.$

Substitution of this value for x into either of the two equations gives $y = 2$.

N. B. When the form of the simultaneous equations is as above, it is customary to include the value of y, as well as that of x, as part of the solution.

Example

Solve $\begin{cases} 2y - 2x = -1 \\ 3y + 4x = 2. \end{cases}$

These have to be first manipulated as follows:

$$\begin{cases} 2y = 2x - 1 \\ 3y = -4x + 2 \end{cases}$$

and hence $\begin{cases} y = x - \frac{1}{2} \\ y = -\frac{4}{3}x + \frac{2}{3}. \end{cases}$

Having done this manipulation we can proceed as before to equate the right-hand sides of both equations, and obtain the solution $x = \frac{1}{2}$, $y = 0$.

4.8.6 Exercises

1. Solve $\begin{cases} y - 2x - 4 = 0 \\ y - 3x - 2 = 0 \end{cases}$

2. Solve $\begin{cases} x^2 + 2x - 1 - y = 0 \\ 2x + 10 - 2y = 0. \end{cases}$

4.8.7 Simultaneous equations in more than two unknowns

We have, so far, only discussed the method of solving two simultaneous equations, but frequently in economics we encounter situations where we have three equations to be solved simultaneously. We shall illustrate by examples how the method suggested above can be generalised to these larger systems of equations.

Example
Solve simultaneously
$$\begin{cases} x + y + z = 2 \\ 2x - y + z = 2 \\ x + 2y - z = 5. \end{cases}$$

What we are required to do is to find values for x, y and z which will simultaneously satisfy *all* three equations. For example, $x = 4$, $y = 2$, $z = -4$ will *not* suffice because although the first two equations would be satisfied the last would not.

In order to solve the problem we shall attempt to reduce it from a problem involving three simultaneous equations in three unknowns to one involving only two simultaneous equations in two unknowns, which we can then solve as in sections 4.8.3 or 4.8.5. If we are to reduce the problem from three unknowns x, y and z to a problem in only two we must 'eliminate' one of the unknowns. Let us choose z to be eliminated. (The choice of which unknown to eliminate is arbitrary.)

We begin by manipulating *each* equation, using the rules of section 4.2.2, until we have z on its own on the left-hand side of each equation. So we get
$$\begin{cases} z = 2 - x - y \\ z = 2 - 2x + y \\ z = -5 + x + 2y. \end{cases}$$

We now equate the right-hand sides of the first two equations to get
$$2 - x - y = 2 - 2x + y.$$
Similarly, equating the right-hand sides of the last two equations we get
$$2 - 2x + y = -5 + x + 2y.$$

We have now got just *two* equations in two unknowns x and y. (We say 'z has been eliminated'.)

Manipulating each of these two equations in turn we can write them as
$$\begin{cases} x - 2y = 0 \\ -3x - y + 7 = 0 \end{cases}$$
and we are left with a problem identical to those discussed in section 4.8.5.

We proceed to 'eliminate y' by first re-writing the two equations so that y is on its own on the left-hand side of both
$$\begin{cases} y = \tfrac{1}{2}x \\ y = 7 - 3x \end{cases}$$
and then equating the right-hand sides gives
$$\tfrac{1}{2}x = 7 - 3x.$$
Therefore $3\tfrac{1}{2}x = 7$.
Therefore $x = 2$.

Substitution of $x = 2$ into any one of the pair of equations involving x and y only, gives $y = 1$ and substitution of both $x = 2$ and $y = 1$ into any one of the original three equations gives $z = -1$.

The solution of the three simultaneous equations is then $x = 2$, $y = 1$, $z = -1$ and substitution of these values into each of the equations confirms this.

In principle this method of repeated elimination can be applied to systems of four equations in four unknowns or even larger systems.

4.8.8 Exercises

Solve the following simultaneous equations

1. $\begin{cases} x - y + z = 7 \\ x + 2y - z = -1 \\ x - 2y + z = 1 \end{cases}$

2. $\begin{cases} 2x + y - 2z = -3 \\ x - y - z = -3 \\ x + y + 2z = 5. \end{cases}$

4.9 Applications to economics

In this section we shall present examples of the use of simultaneous equations, but we shall not give examples of the use of a single equation. The reader will find that the use of single equations is necessary for many of the economic examples presented later in the book, e.g., in section 6.6, and indeed in this section, since solving simultaneous equations algebraically, involves solving single equations.

4.9.1 Equilibrium in a simple market model

We have already introduced the reader to the idea of a demand function (section 2.10.1) relating the quantity of a good demanded, q^d, to the price, p, of the good, according to some function, $f(p)$.
For example, $q^d = 30 - 2p$.
Similarly, we might have a supply function for the same good relating the quantity supplied, q^s, to the price, p, according to some function, $g(p)$ (see section 2.10.4).
For example, $q^s = 5 + 3p$.

The market is said to be in 'equilibrium' when the quantity demanded, q^d, is exactly equal to the quantity supplied, q^s. We might then ask what the price must be in order for the market to be in equilibrium. That is, we want a value for p such that the quantity demanded, which is given by $f(p)$, is equal to the quantity supplied, which is given by $g(p)$. This is exactly the simultaneous equation problem described in section 4.8.1, and to find the equilibrium value of p we adopt the method of section 4.8.3.
Example
Given the following demand and supply functions for a particular good, find the equilibrium price (i.e. the value of p which makes the quantity demanded equal to that supplied).
$q^d = 30 - 2p$
$q^s = 5 + 3p$.
We require a value of p such that $30 - 2p = 5 + 3p$,
i.e. we require p such that $25 = 5p$.
So the equilibrium price is $p = 5$.

Substituting this value for p into the demand function gives $q^d = 20$, and substituting into the supply function gives $q^s = 20$, i.e. the quantity demanded *is* equal to that supplied.

The equilibrium value for p can alternatively be found by the graph method of section 4.8.2.

The demand function and supply function are both drawn on the same diagram (Fig. 4.15).

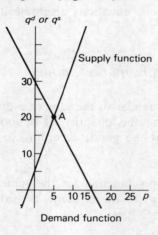

Fig. 4.15

The graphs intersect at the point A which corresponds to the ordered pair $(5, 20)$. The first element of this pair gives us the solution to the simultaneous equation, i.e. the equilibrium price, $p = 5$.

N. B. The second element of the ordered pair gives us the equilibrium quantity. That is, the quantity both demanded and supplied at the equilibrium price.

Remark

The demand and supply graphs in this book are drawn with price p represented by the *horizontal* axis. It is common to find in economics texts that the axes are reversed.

4.9.2 *The imposition of a tax on the simple model*

In this section we retain the demand and supply functions used in section 4.9.1 except that we shall introduce a tax equal to £10, on every item sold. The easiest way to incorporate this is into the supply function. The quantity supplied by producers, q^s, is related to the price they *receive*, which is no longer the same as the price the consumers pay.

If we denote the price the suppliers receive by p^*, our supply function should be written in terms of p^*, i.e. $q^s = 5 + 3p^*$.

If we retain the use of p to denote the price the consumer pays then we have $p = p^* + 10$, i.e. for each item purchased the consumer has to pay the supplier p^* and in addition a tax of £10. If $p = p^* + 10$ then $p^* = p - 10$ and the supply function can be written in terms of p as

$$q^s = 5 + 3(p - 10)$$
$$= 5 + 3p - 30.$$

Therefore $q^s = -25 + 3p$.

The demand function is already given in terms of p as

$q^d = 30 - 2p$.

We can now proceed as before to determine the new equilibrium in our model with a tax.

$q^d = 30 - 2p$
$q^s = -25 + 3p$

and we require a value of p such that

$30 - 2p = -25 + 3p$,

i.e. we require p such that $55 = 5p$.

So the new equilibrium price is $p = 11$.

N. B. Notice how the introduction of a tax has produced an increase in the equilibrium price from $p = 5$ to $p = 11$.

There will, of course, be a corresponding fall in the quantity demanded and supplied, from 20 to 8.

In diagrammatic terms we have that the supply function has shifted down (Fig. 4.16). (The intercept has changed from 5 to -25.)

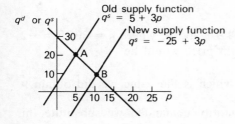

Fig. 4.16

The old equilibrium point was at A corresponding to the ordered pair (5, 20), and the new equilibrium point is at B corresponding to the pair (11, 8).

4.9.3 General linear market model

The above examples were of the market for a particular good, using particular demand and supply functions. In this example we generalise by not specifying particular demand and supply functions although we restrict the example to demand and supply functions which are linear. (The advantage of such a general approach is that the results established in this section (and later in section 4.9.4), apply to *any* market which has linear demand and supply functions.)

The demand function is $q^d = a + bp$ where a and b can be *any* numbers providing only that $a > 0$ and $b < 0$. ($b < 0$ ensures that the quantity demanded decreases as price increases, $a > 0$ ensures that the quantity demanded is a positive number.)

The supply function is $q^s = c + dp$ where c and d can be *any* numbers providing only that $d > 0$.

(The example in section 4.9.1 above, is a particular case of this more general example, with $a = 30$, $b = -2$, $c = 5$, $d = 3$.)

To find the equilibrium price we require a value of p such that

$$a + bp = c + dp,$$

i.e. such that
$$a - c = dp - bp$$
$$a - c = (d - b)p$$
$$\frac{a - c}{d - b} = p.$$

So the equilibrium price is given by

$$p = \frac{a - c}{d - b}.$$

(In the example in section 4.9.1 above, where $a = 30$, $b = -2$, $c = 5$, $d = 3$, the equilibrium price would be

$$p = \frac{30 - 5}{3 - (-2)} = \frac{25}{5} = 5,$$ which is the value we did obtain.)

To find the equilibrium quantity demanded we substitute this value for p into the demand function to get

$$q^d = a + bp$$

$$= a + b\left(\frac{a-c}{d-b}\right)$$

$$= \frac{a(d-b) + b(a-c)}{d-b}$$

$$= \frac{ad - ab + ba - bc}{d-b},$$

i.e. $q^d = \frac{ad - bc}{d-b}$.

Substituting the equilibrium value of p into the supply function shows that q^s is also equal to this.

Remark
We have already placed certain restrictions on the values that a, b, and d can take, if the demand and supply functions are to be realistic; namely that $a > 0$, $b < 0$ and $d > 0$. We need an additional restriction for our equilibrium to be realistic; namely that $a > c$, i.e. that the intercept of the demand function be greater than that of the supply function. This extra condition guarantees that the equilibrium price will be positive as shown below.

The equilibrium price $p = \frac{a-c}{d-b}$.

Since $d > 0$ and $b < 0$ we know that the denominator of this expression will be > 0. If $a > c$ then we also know that $a - c > 0$ (using the rule for manipulating inequalities). Since both the numerator and denominator of the expression are >0 we can conclude that the whole expression is >0.

4.9.4 The imposition of a tax on the general model

In this example we retain the general demand and supply functions used in section 4.9.3, but introduce a tax as we did for the particular model in section 4.9.2.

Suppose a tax equal to an amount £t on every item sold, is introduced. As we saw in section 4.9.2, this is easiest to incorporate into the supply function. If we denote the price the producer *receives* by p^*, then the supply function should be written in terms of p^* as
$$q^s = c + dp^*.$$
p^* is related to the price, p, the consumers pay, by
$$p = p^* + t$$
or $\quad p^* = p - t$
and hence the supply function can be written in terms of p as
$$q^s = c + d\,(p - t)$$
or $\quad q^s = (c - dt) + \mathrm{d}p.$
The demand function is already given in terms of p as
$$q^d = a + bp.$$
We can now proceed as before to find the equilibrium price as being the value of p such that
$$a + bp = (c - dt) + dp$$
$$a - c + dt = dp - bp$$
$$a - c + dt = (d - b)p.$$
So the equilibrium price is given by

$$p = \frac{a - c + dt}{d - b}.$$

(In the example in section 4.9.2 where $a = 30$, $b = -2$, $c = 5$ and $d = 3$, we imposed a tax of $t = 10$, and the new equilibrium price was

$$p = \frac{30 - 5 + 3 \cdot 10}{3 - (-2)} = \frac{55}{5} = 11.)$$

We can re-write the formula for the equilibrium price as

$$p = \frac{a - c}{d - b} + \frac{dt}{d - b}$$

and observe that this equilibrium price is made up of two components. The first part, $(a - c)/(d - b)$, is the equilibrium price which *would* have prevailed if there was no tax (see section 4.9.3). But in a market where there is a tax the price is *increased* over this value by an amount $dt/(d - b)$.

N. B. It is an *increase* since $d > 0$, $d - b > 0$ and $t > 0$, hence the whole expression is >0.

N. B. The increase in price is *less* than the amount of tax, t, since

$$\frac{d}{d-b} < 1, \text{ if } b < 0.$$

The example can again be considered in terms of a graph diagram (Fig. 4.17).

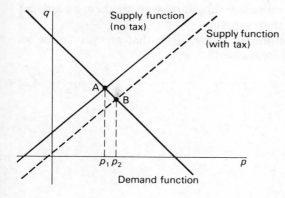

Fig. 4.17

In the diagram there are two supply functions drawn, one corresponding to the model with no tax, and another *below* the first one, corresponding to the model with a tax. (The 'with tax' supply function has an intercept $c - dt$, which if $d > 0$ and $t > 0$ is smaller than that of the 'without tax' supply function.)

In the 'without tax' model the equilibrium price, p_1, is obtained from the point A (where the demand and supply function intersect). In the 'with tax' model the equilibrium price, p_2, is obtained from point B (where the demand and *new* supply function intersect). It is easily seen that $p_2 > p_1$.

Remark

The quantity demanded and supplied at the equilibrium price can be derived by substituting the equilibrium price into either the demand or supply function to give

$$q^d = q^s = \frac{da - bc}{d - b} + \frac{bdt}{d - b}.$$

Remark
A similar analysis can be performed on simple macro-economic models and the reader is referred to Archibald and Lipsey *A Mathematical Treatment of Economics*.

4.9.5 Exercises

Using the simple market model presented in section 4.9.3 above, what is the effect on the equilibrium price if:
1. The government *subsidises* suppliers by giving them an amount s, for each unit they supply?
2. The tastes of consumers change such that at every price an extra x units are demanded?

Chapter 5

Differentiation of functions of one variable

In Chapter 2 we defined cost and revenue functions for a firm producing a single commodity and pointed out that as output changed so the costs (and revenue) for the firm would change. In Chapter 4 we considered simple market models and saw how the equilibrium price (and quantity) changed if taxes or subsidies were changed. The concept of 'change' in variables is very important in economics and the technique frequently employed to consider the change is 'differentiation'. In this chapter we define differentiation and outline the methods for differentiating and hence for finding the way in which variables change. In section 5.6 we apply the techniques of differentiation to analyse the changes in cost and revenue as output changes, and the changes in equilibrium price (and quantity) as tax rates change.

5.1 Introduction

In section 2.7.1 we discussed linear functions, i.e. functions of the form
$f(x) = ax + b$,
and we stated that such functions had graphs which were straight lines, with slopes given by the number 'a'.
Example
$f(x) = 2x + 1$ has the following graph (Fig. 5.1). The slope of the graph is 2, i.e. for every 1 unit we move to the right we have to move 2 units up, to stay on the graph.
Example
$f(x) = -3x - 1$ has the graph shown in Fig. 5.2. The slope is -3, i.e. for every 1 unit we move to the right we have to move -3 units up, or $+3$ units *down*, to stay on the graph.
   Clearly the slope of the graph of a *linear* function is well defined.

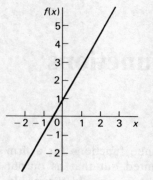

Fig. 5.1

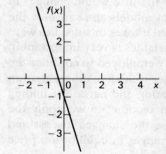

Fig. 5.2

N. B. The bigger the slope, the faster the graph rises. A negative slope indicates a falling graph.

In this chapter we ask ourselves 'is it reasonable to talk about the slope of the graph of a function, even if the function is not linear?' Our first reaction might be that it is straightforward. For example, given the graphs of two functions, $f(x)$ and $g(x)$ (Fig. 5.3), it seems reasonable to say that the graph of $f(x)$ is steeper than that of $g(x)$.

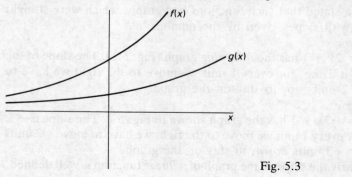

Fig. 5.3

However, given the graph of a function $h(x)$ (Fig. 5.4), the problem appears less simple.

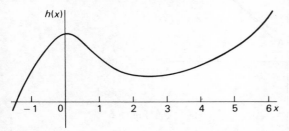

Fig. 5.4

Here the slope (or steepness) of $h(x)$ varies. When x is approximately 1, the graph is falling (i.e. it has a negative slope), but when x is about 4 the graph is rising (i.e. the slope is positive) and when $x = 6$ the graph is even steeper (i.e. the slope is bigger). In general, for most functions (linear functions are an exception) the slope of the graph of the function will be different at different places on the graph, and for this reason we cannot talk about *the slope of the graph of the function*, only about *the slope of the graph of the function at a point*. (In fact we shall discover that the slope is given by another function of x.)

Since we have no such problems with the slopes of linear functions, we define the slopes of all other functions in terms of slopes of linear functions.

5.2 The slope of a function at a point

Definition
We define the slope of a function $f(x)$, at a point x^*, to be the slope of the tangent to the graph of the function at x^*. In our diagram (Fig. 5.5) the tangent to the graph of the function, at x^* is the line AB. Since the tangent is linear it has a well defined slope.
N. B. At different points, the tangents to the curve will vary and hence the slope of the function varies at different points. In Fig. 5.6, the tangent at x_1 is much steeper than at x_2 (i.e. has a greater slope), therefore, the slope of $f(x)$ is greater at x_1 than at x_2. The tangent at x_3 has a negative slope and hence the slope of $f(x)$ at x_3, is negative.

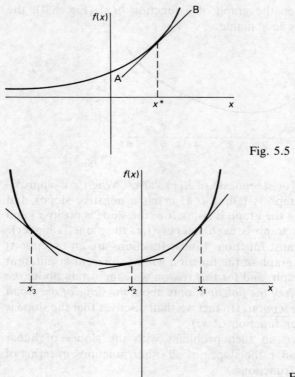

Fig. 5.5

Fig. 5.6

We now know what we mean by the slope of a function, and now consider how we would calculate it given a particular function. Clearly drawing the graph (it would have to be very accurate), drawing a tangent (again it would need to be accurate) and working out the slope of the tangent, would be a laborious method, and mathematicians have developed a technique which is far quicker – namely *differentiation*.

5.3 A method for finding the slope of a function

In this section we provide an explanation of the method mathematicians have used to calculate the slope of virtually any function at any point. In practice this method is *not* used – a set of rules, which we shall give in section 5.5, enables us to calculate slopes far more easily. However, this section may be of interest to students, although it is not essential for subsequent developments.

Suppose we wish to find the slope of the graph of a function $f(x)$ at a point x^* (Fig. 5.7).

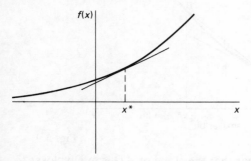

Fig. 5.7

Consider first a value of x a bit bigger than x^*, let's call it $x^* +$ a bit.

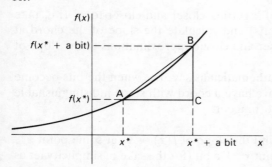

Fig. 5.8

We can find the images of both x^* and the new value, $x^* +$ a bit, under the function f (i.e. $f(x^*)$ and $f(x^* +$ a bit)) and hence we can find the slope of the *chord AB* (Fig. 5.8).

N. B. The diagram exaggerates the size of 'a bit' to improve the clarity.

The slope of the chord AB is given by

$$\text{slope} = \frac{BC}{AC} = \frac{f(x^* + \text{a bit}) - f(x^*)}{(x^* + \text{a bit}) - x^*}$$

$$= \frac{f(x^* + \text{a bit}) - f(x^*)}{\text{a bit}}.$$

If we now take another value of x, bigger than x^*, but smaller than $x^* +$ a bit — let's call it $x^* +$ a smaller bit, we can in a similar way calculate the slope of the new chord – AD in the diagram (Fig. 5.9).

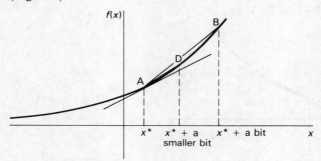

Fig. 5.9

From the diagram it is clear that the slope of AD is much closer to the slope of the tangent (which is the thing we are interested in), than the slope of AB.

If we take values of x getting closer and closer to x^* (i.e. take smaller and smaller bits) and calculate the slope of the chord at each step, we get closer and closer approximations to the slope of the tangent.

'In the limit', as mathematicians say, i.e. when the bits become infinitesimally small, we have a chord with slope indistinguishable from the slope of the tangent.

Example

Suppose we wish to find the slope of $f(x) = x^2$ at some point x^*. We consider a new point $x^* +$ a bit (for the sake of simplicity let us call it $x^* + h$). The slope of the chord AB is then

$$\frac{f(x^* + h) - f(x^*)}{h} = \frac{(x^* + h)^2 - (x^*)^2}{h}.$$

So, expanding, we get the slope of AB as

$$\frac{(x^* + h)(x^* + h) - x^{*2}}{h} = \frac{x^{*2} + 2x^*h + h^2 - x^{*2}}{h}$$

$$= \frac{2x^*h + h^2}{h}$$

$$= 2x^* + h.$$

We now take values closer and closer to x^* (i.e. take smaller bits or smaller values for h). At each step the slope of the chord will be $2x^* + h$ (but remember h is getting smaller). 'In the limit', i.e. as h gets infinitesimally small, this becomes $2x^*$ – and this is the slope of the tangent, and hence the slope of the graph of the function at x^*.

N. B. We have calculated the slope at x^* without actually specifying what x^* is, i.e. we have a formula $2x^*$, which gives us the slope of the graph of the function at *any* point. That is, the slope of $f(x) = x^2$ at any value x^* is given by $2x^*$. In fact since x^* can be any value it is more convenient to label it by our conventional x.

So, the slope of $f(x) = x^2$, at any value of x, is given by $2x$.
So, for example, the slope when $x = 3$ is $2 \cdot 3 = 6$.
The slope when $x = -1$ is -2, and the slope when $x = 0$ is zero.

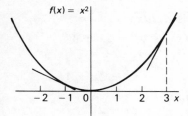

Fig. 5.10

These values do appear to be consistent with our diagram (Fig. 5.10).

5.4 Derivatives and differentiation

Definition
In general, given almost any function $f(x)$ it is possible, using the method above, to find the slope of $f(x)$ at *any* point. These slopes will be given by a formula or, if you like, another function (in the example above $2x$). This other function is called the *derivative of* $f(x)$ and denoted by $f'(x)$.
For example, if $\qquad f(x) = x^2$
then the derivative $\quad f'(x) = 2x$.

Definition
We say we *differentiate $f(x)$* to get the *derivative $f'(x)$*.

Given virtually any function $f(x)$ it is possible to find the derivative $f'(x)$ and hence the slope of $f(x)$.

Example

If $f(x) = x^2 - 2x + 4$
differentiation gives $f'(x) = 2x - 2$.
(We shall see in the next section how this derivative was obtained.)
Hence the slope of $f(x)$ when $x = 3$ is obtained by substituting 3 into the $f'(x)$ function. Therefore the slope is 4.
Similarly, when $x = -1$ the slope of $f(x)$ is $2 \cdot (-1) - 2$, i.e. -4.

Remark

The derivative $f'(x)$ gives the slope of $f(x)$, i.e. it tells us how $f(x)$ is changing and is often referred to as the *rate of change* of $f(x)$.

Remark

An alternative notation for $f'(x)$ is dy/dx.
So, for example, instead of 'if $f(x) = x^2 - 2x + 4$
then $f'(x) = 2x - 2$'
you might find written 'if $y = x^2 - 2x + 4$
then $\dfrac{dy}{dx} = 2x - 2$'.

Remark

Not all functions can be differentiated. A function can be differentiated only if it is *continuous*. A function is continuous if it has no breaks or jumps in its graph. (The formal definition of continuity need not concern us here and the interested reader is referred to, for example, Yamane *Mathematics for Economists*.) The functions illustrated below are not continuous.

(a)

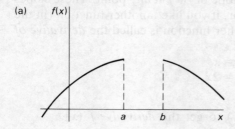

This function is not defined for values of x between a and b.

(b)
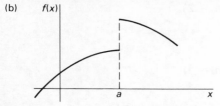

This function has a 'jump' discontinuity.

(c)
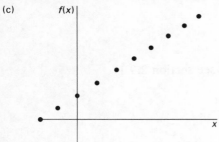

This is a *discrete* function only defined when x is an integer.

The problem with a function which is not continuous is that it is impossible to define the slope of the function at the points of discontinuity.

5.5 Rules of differentiation

The method outlined in section 5.3 for finding the slope of a function is obviously very tedious, and rather than adopt this method every time, we shall introduce a *set of rules* which enable us to differentiate almost any function much more quickly. The proofs that these rules are valid are based on the method of section 5.3 but we shall not prove them here. (Interested readers may consult, for example, Parry Lewis *An Introduction to Mathematics for Students of Economics* or Yamane *Mathematics for Economists*.)

5.5.1 The power rule

If $f(x) = ax^n$ where a and n are *any* two numbers
then $f'(x) = a \cdot n \cdot x^{n-1}$
(We say we 'bring down the power and reduce the power by one'.)

Example
If $f(x) = 3x^2$ $(a = 3, n = 2)$
then $f'(x) = 3 \cdot 2 \cdot x^1$
$= 6x$.

Example
If $f(x) = 2x^5$ $(a = 2, n = 5)$
then $f'(x) = 2 \cdot 5 \cdot x^4$
$= 10x^4$.

Example
If $f(x) = -2x^3$
then $f'(x) = -6x^2$.

Example
If $f(x) = \dfrac{4}{x} = 4 \cdot x^{-1}$ (see section 2.9.1)

then $f'(x) = -4x^{-2} = \dfrac{-4}{x^2}$.

Example
If $f(x) = \dfrac{2}{x^3} = 2 \cdot x^{-3}$

then $f'(x) = -6x^{-4} = \dfrac{-6}{x^4}$.

Special case
If $f(x) = x^3$ (here $a = 1$)
then $f'(x) = 3x^2$.

Special case
If $f(x) = 4x$ (here $n = 1$, since $x = x^1$).
then $f'(x) = 4 \cdot 1 \cdot x^0 = 4x^0$
$= 4$ (since $x^0 = 1$).
Generally if $f(x) = ax$, for any number a, then $f'(x) = a$.

Special case
$f(x) = 7$
We can treat this as $f(x) = 7 \cdot x^0$ (since $x^0 = 1$),
and then $f'(x) = 7 \cdot 0 \cdot x^{-1} = 0$.
Generally if $f(x) = a$, for any number a, then $f'(x) = 0$, i.e. the derivative of a constant is zero.

5.5.2 Differentiation of sums

If $f(x) = g(x) + h(x)$ for two functions g and h, then $f'(x) = g'(x) + h'(x)$.

Example
If $f(x) = 3x^2 + 2x^5$
then $f'(x) = 6x + 10x^4$.
(We say we differentiate 'term by term'.)

Example
If $f(x) = x^2 + 5x$
then $f'(x) = 2x + 5$.

The rule extends naturally to differentiating a sum of *three* (or more) terms.

Example
If $f(x) = 3x^3 + 4x^2 + 7x$
then differentiating 'term by term'
$f'(x) = 9x^2 + 8x + 7$.

Example
If $f(x) = x^3 + 3x^2 + 4x + 5$
then $f'(x) = 3x^2 + 6x + 4$.

5.5.3 Differentiation of differences

This rule is very similar to that for sums, given in section 5.5.2.
If $f(x) = g(x) - h(x)$, then $f'(x) = g'(x) - h'(x)$.

Example
If $f(x) = 2x^3 - 4x^5$
then $f'(x) = 6x^2 - 20x^4$.

The rule can be extended to cover more than two terms and can be combined with section 5.5.2.

Example
If $f(x) = x^2 - 4x - 3$
then $f'(x) = 2x - 4$.

Example
If $f(x) = x^3 - 4x^2 + 7x - 6$
then $f'(x) = 3x^2 - 8x + 7$.

5.5.4 Exercises

Differentiate the following functions, and hence find the slopes of the functions at the points $x = 0$, $x = 1$, $x = -1$.

1. $f(x) = x^2 - 3x + 2$
2. $f(x) = 1 - x - 4x^2 + x^3$
3. $f(x) = 3x + 7$.

5.5.5 Differentiation of products

It would be very convenient if we could have a rule similar to the above for functions of the form $f(x) = g(x) \cdot h(x)$.
Unfortunately, it is *not* true that $f'(x) = g'(x) \cdot h'(x)$. Indeed, the correct rule is rather more complicated. The rule is as follows.
If $\quad f(x) = g(x) \cdot h(x)$
then $\quad f'(x) = g'(x) \cdot h(x) + h'(x) \cdot g(x)$.
Example
If $f(x) = (3x + 2) \cdot (2x - 1)$
then $\quad g(x) = 3x + 2 \quad$ and $\quad h(x) = 2x - 1$
and so $\quad g'(x) = 3 \quad\quad\quad$ and $\quad h'(x) = 2$.
Using our rule in this section 5.5.5 we can write
$$f'(x) = g'(x) \cdot h(x) + h'(x) \cdot g(x)$$
$$= (3) \cdot (2x - 1) + (2) \cdot (3x + 2)$$
$$= (6x - 3) + (6x + 4).$$
Therefore $f'(x) = 12x + 1$.
Example
If $f(x) = (1 - 7x) \cdot (x^2 + 2x - 5)$
then $\quad g(x) = 1 - 7x \quad$ and $\quad h(x) = x^2 + 2x - 5$
and so $\quad g'(x) = -7 \quad\quad$ and $\quad h'(x) = 2x + 2$.
Using our products rule we have
$$f'(x) = g'(x) \cdot h(x) + h'(x) \cdot g(x)$$
$$= (-7) \cdot (x^2 + 2x - 5) + (2x + 2) \cdot (1 - 7x)$$
$$= (-7x^2 - 14x + 35) + (2x - 14x^2 + 2 - 14x)$$
$f'(x) = -21x^2 - 26x + 37$.

Special case
If $f(x) = c \cdot h(x)$ where c is any constant, then using the products rule with $g(x) = c$, we have
$$f'(x) = g'(x) \cdot h(x) + h'(x) \cdot g(x)$$
$$= 0 \cdot h(x) + h'(x) \cdot c$$
i.e. $f'(x) = c \cdot h'(x)$.
Example
If $\quad f(x) = 3 \cdot (4x^2 + 2x - 3)$
then $\quad f'(x) = 3 \cdot (8x + 2)$.

We shall refer to this special case as the 'constant multiple' rule of differentiation.

5.5.6 Differentiation of quotients

The rule for functions of the form $g(x)/h(x)$ is even more complicated. It is:

If $f(x) = \dfrac{g(x)}{h(x)}$

then $f'(x) = \dfrac{g'(x) \cdot h(x) - h'(x) \cdot g(x)}{(h(x))^2}$.

N. B. The top line is *almost* identical to that in the expression for the products rule, *except for the minus sign.*
In addition we divide by the square of the function $h(x)$.

Example

If $f(x) = \dfrac{2x - 5}{4x + 2}$

then $g(x) = 2x - 5$ and $h(x) = 4x + 2$
and so $g'(x) = 2$ and $h'(x) = 4$.
Therefore, using the quotients rule,

$$f'(x) = \frac{g'(x) \cdot h(x) - h'(x) \cdot g(x)}{(h(x))^2}$$

$$= \frac{(2) \cdot (4x + 2) - (4) \cdot (2x - 5)}{(4x + 2)^2}$$

$$= \frac{(8x + 4) - (8x - 20)}{(4x + 2)^2}$$

(Notice that when the brackets are removed from the second expression *all* the signs are changed.)

$$= \frac{8x + 4 - 8x + 20}{(4x + 2)^2}$$

$$f'(x) = \frac{24}{(4x + 2)^2}.$$

Example

If $f(x) = \dfrac{4x^2 - 3}{2x}$

then $g(x) = 4x^2 - 3$ and $h(x) = 2x$
and so $g'(x) = 8x$ and $h'(x) = 2$.
Therefore, using the rule

$$f'(x) = \frac{g'(x) \cdot h(x) - h'(x) \cdot g(x)}{(h(x))^2}$$

$$= \frac{(8x) \cdot (2x) - (2) \cdot (4x^2 - 3)}{(2x)^2}$$

$$= \frac{(16x^2) - (8x^2 - 6)}{4x^2}$$

$$= \frac{8x^2 + 6}{4x^2}.$$

With care, we can simplify by dividing the top (the *whole* of the expression) and the bottom by 2 to give:

$$f'(x) = \frac{4x^2 + 3}{2x^2}.$$

5.5.7 Exercises

Differentiate the following functions using the rules in sections 5.5.5 and 5.5.6.

1. $f(x) = (1 + x^2) \cdot (3x - 2)$
2. $f(x) = x^3 \cdot (1 - x)$
3. $f(x) = \dfrac{x^2}{1 - x}$
4. $f(x) = \dfrac{x^2 + 2}{1 - 3x}.$

5.5.8 Function of a function rule

We have seen in the previous sections how to differentiate $g(x) + h(x), g(x) - h(x), g(x) \cdot h(x)$ and $g(x)/h(x)$. If you refer back to Chapter 3, in particular to section 3.3, you will find that there is one way of combining functions for which we do not yet have a rule of differentiation.

Example
If $g(x) = x^3$ and $h(x) = 2x + 1$
we can form a function $f(x) = g(h(x))$.
We require $\qquad g(2x + 1)$
and we know that $g(x) = x^3$
or g (any number) = (that number)^3.
Therefore $g(2x + 1) = (2x + 1)^3$,
i.e. $\qquad g(h(x)) = (2x + 1)^3$.
So $\qquad\qquad f(x) = (2x + 1)^3$.

We now ask 'is there a rule which will enable us to get the derivative of $f(x)$, knowing the derivatives of the constituent functions $g(x)$ and $h(x)$?'
This rule is called the *function of a function* rule of differentiation. The rule is:
$\qquad$ If $f(x) = g(h(x))$
then $\quad f'(x) = g'(h(x)) \cdot h'(x)$.

Example
In the example above $f(x) = (2x + 1)^3$ can be written $f(x) = g(h(x))$ where $h(x) = 2x + 1$
and $g(x) = x^3$.
So $h'(x) = 2$ and $g'(x) = 3x^2$.
Then using the function of a function rule
$\qquad f'(x) = g'(h(x)) \cdot h'(x)$
$\qquad\qquad = 3(2x + 1)^2 \cdot 2$,
i.e. $f'(x) = 6(2x + 1)^2$.

Example
$\qquad$ If $f(x) = (1 - 3x^2)^5$
then $\qquad f(x) = g(h(x))$ where $h(x) = 1 - 3x^2$ and $g(x) = x^5$.
So $h'(x) = -6x$ and $g'(x) = 5x^4$.
Then using the function of a function rule
$\qquad f'(x) = g'(h(x)) \cdot h'(x)$
$\qquad\qquad = 5(1 - 3x^2)^4 \cdot -6x$,
i.e. $f'(x) = -30x(1 - 3x^2)^4$.

N. B. When breaking down $f(x)$ into $g(h(x))$, the $h(x)$ function is the part in brackets.

Example
$\qquad$ If $f(x) = (2 - x^5)^2$
then $\qquad f(x) = g(h(x))$ where $h(x) = 2 - x^5$ and $g(x) = x^2$.
So $h'(x) = -5x^4$ and $g'(x) = 2x$.

Using the function of a function rule
$$f'(x) = g'(h(x)) \cdot h'(x)$$
$$= 2(2 - x^5) \cdot -5x^4$$
$$f'(x) = -10x^4(2 - x^5).$$
N. B. When using the function of a function rule we need $g'(h(x))$, not $g'(x)$, in the formula.

5.5.9 Differentiation of special functions (sin x, cos x, $\log_e x$, e^x)

(a) If $f(x) = e^x$
then $f'(x) = e^x$.

Remark
The exponential function is the only function which has, as its derivative, the same function.

(b) If $f(x) = \log_e x$

then $f'(x) = \dfrac{1}{x}$

(c) If $f(x) = \sin(x)$
then $f'(x) = \cos(x)$
(d) If $f(x) = \cos(x)$
then $f'(x) = -\sin(x).$

N. B. The above rules apply only to the specific functions mentioned, not to hybrids of these functions. However, we can use the above rules together with the function of a function rule to differentiate functions like $f(x) = e^{3x}$ or $f(x) = \sin(3x - 2)$.

Example
If $f(x) = e^{3x}$
then $f(x) = g(h(x))$ where $h(x) = 3x$ and $g(x) = e^x$
and so $h'(x) = 3$ and $g'(x) = e^x$.
Using the function of a function rule
$$f'(x) = g'(h(x)) \cdot h'(x)$$
$$= e^{3x} \cdot 3,$$
i.e. $f'(x) = 3e^{3x}.$

Example
If $f(x) = \log_e(x^2)$

then $f(x) = g(h(x))$ where $h(x) = x^2$ and $g(x) = \log_e x$

and so $h'(x) = 2x$ and $g'(x) = \dfrac{1}{x}$.

Using the function of a function rule
$$f'(x) = g'(h(x)) \cdot h'(x)$$
$$= \dfrac{1}{x^2} \cdot 2x,$$

i.e. $f'(x) = \dfrac{2}{x}$.

Remark
An alternative way of deriving the same result is as follows
$f(x) = \log_e(x^2) = 2 \log_e(x)$
(using property 3 of log functions given in section 2.9.5).
Therefore using the constant multiple rule for differentiation (see section 5.5.5):

$$f'(x) = 2 \cdot \dfrac{1}{x} = \dfrac{2}{x}.$$

Example
If $f(x) = \sin(3x - 4)$
then $f(x) = g(h(x))$ where $h(x) = 3x - 4$ and $g(x) = \sin(x)$
and so $h'(x) = 3$ and $g'(x) = \cos(x)$.
Using the function of a function rule
$$f'(x) = g'(h(x)) \cdot h'(x)$$
$$= \cos(3x - 4) \cdot 3,$$
i.e. $f'(x) = 3 \cos(3x - 4)$.

Example
If $f(x) = (\cos(x))^2$
then $f(x) = g(h(x))$ where $h(x) = \cos(x)$ and $g(x) = x^2$,
and so $g'(x) = 2x$ and $h'(x) = -\sin(x)$.
Using the function of a function rule
$$f'(x) = g'(h(x)) \cdot h'(x)$$
$$= 2 \cos(x) \cdot -\sin(x),$$
i.e. $f'(x) = -2 \cos(x) \sin(x)$.

Remark
$f(x) = (\cos(x))^2$ is not the same as $f(x) = \cos(x^2)$
(the $g(x)$ and $h(x)$ are switched).

Remark
$f(x) = (\cos(x))^2$ is sometimes written as
$f(x) = \cos^2(x)$. Similarly, $f(x) = \sin^2(x)$ means
$f(x) = (\sin(x))^2$ *and not* $f(x) = \sin(x^2)$.

5.5.10 Exercises

1. Differentiate the following functions and hence find the slopes of the functions at the points $x = 0$, $x = 1$.
   (a) $f(x) = e^x$
   (b) $f(x) = \sin(x)$
   (c) $f(x) = \cos(x)$.
2. Find the slope of $f(x) = \log_e(x)$ at $x = 1$. Can you find the slope when $x = 0$?
3. Differentiate the following:
   (a) $f(x) = e^{x^2+1}$
   (b) $f(x) = e^{-3x}$
   (c) $f(x) = \log_e(2x - 1)$
   (d) $f(x) = \sin(1 - 4x)$
   (e) $f(x) = \cos(x^2)$
   (f) $f(x) = \sin^2 x + \cos^2 x$.

5.6 Applications to economics

5.6.1 Marginal cost

In section 2.10.6 we introduced the idea of a *cost function* for a firm, which gave the total cost incurred by the firm in producing a certain level of output, as a function of that level of output.
Example
A firm has a cost function given by:
  cost = $x^3 - 12x^2 + 60x + 20$.

We also defined the *average cost*, to be the cost/unit of output. So in the example above,

$$\text{average cost} = \frac{\text{cost}}{\text{output}} = \frac{x^3 - 12x^2 + 60x + 20}{x}$$

$$= x^2 - 12x + 60 + \frac{20}{x}.$$

Again we have that average cost is a function of output.

We now introduce a new idea, that of the *marginal cost*.

Definition
For a firm with a total cost function $f(x)$, the *marginal cost* of the firm, at any output, is given by $f'(x)$.

Example
A firm has a total cost function given by:
  cost $= x^3 - 12x^2 + 60x + 20$.
Then the marginal cost function is given by:
  marginal cost $= 3x^2 - 24x + 60$.

Remark
The marginal cost is a function of x and hence we can calculate the marginal cost at *any* output. In the example above, if output $x = 2$ then the marginal cost would be 24 (by substitution into the marginal cost function). The following table gives some of the ordered pairs:

Output x	Marginal cost
0	60
2	24
5	15
10	120

Remark
The marginal cost (unlike the total cost) need not necessarily increase as output increases.

Remark
If the total cost function cannot be differentiated then clearly the above definition of marginal cost will break down.

Remark

The marginal cost at any level of output gives an indication of how costs will change if output is slightly increased or decreased.

In fact the marginal cost at a given level of output, is sometimes defined as the increase in cost that will be incurred if one more unit of the good is produced. This definition is not quite the same as ours, since in using the concept of a derivative, we are dealing with very minute increases in output rather than an increase of one whole unit. For the purposes of this book our definition is preferable but there are cases when the so-called 'discrete' definition will be superior.

5.6.2 Exercise

Given the following total cost function for a firm, find the marginal cost formula and average cost formula, and hence find the total, marginal, and average costs if output is:
 (a) 1   (b) 2   (c) 10.
Total cost = $x^3 - 4x^2 + 10x + 200$.

5.6.3 Marginal revenue

Analogous to the concept of total and marginal costs we have total and marginal revenue, where our definition of marginal revenue is as follows.

Definition

For a firm with a total revenue function $g(x)$, the *marginal revenue* of the firm, at any output, is given by $g'(x)$.

Example

A firm has a total revenue function given by
 revenue = $100x - 4x^2$.
Then the marginal revenue function is given by
 marginal revenue = $100 - 8x$.

Remark

The marginal revenue is a function of x and hence we can calculate the marginal revenue at *any* level of output, as the examples in the following table illustrate.

Output x	Marginal revenue
0	100
2	84
5	60
20	−60

Notice that marginal revenue can be negative. This means that at this level of output any increase in output will *reduce* total revenue.

Remark

If the total revenue function cannot be differentiated, then clearly this definition of marginal revenue breaks down.

Remark

The marginal revenue, at any level of output, gives an indication of how revenue will change if output is slightly increased or decreased. The marginal revenue at a given level of output is sometimes defined as the increase in revenue the firm will receive if one more unit of the good is produced and sold. As with the 'discrete' definition for marginal cost (section 5.6.1), this is not quite the same as our definition involving the derivative.

Remark

The marginal revenue need *not* be the same as the price as the following example shows.

Example

A firm is faced with the demand function $p = 100 - 4x$. If the firm produced 5 units then the price will be $p = 100 - 4 \cdot 5 = 80$.

The total revenue $= p \cdot x = (100 - 4x)x$
$= 100x - 4x^2$

and hence the marginal revenue is given by

marginal revenue $= 100 - 8x$.

When $x = 5$, the marginal revenue $= 100 - 40 = 60$, compared with the price of 80.

Marginal revenue will be the same as the price only if the firm's price does not depend on x, e.g. in a perfectly competitive market where the output of the firm is an insignificant part of the total market supply, and changes in the firm's output have no effect on the market. In this case marginal revenue does not depend on output x and the marginal revenue function is a *constant function* (see section 2.7.2).

5.6.4 Exercise

A firm is faced with the following demand function for its product
$p = 50 - 5x$.
Find the total revenue and marginal revenue functions. Hence find the price, total revenue and marginal revenue at the following outputs:
 (a) 0   (b) 2   (c) 5   (d) 10.

5.6.5 The effect of tax changes on a simple market model

In section 4.9 we considered the effect of the imposition of a tax on a simple market model.

The model consisted of a demand function and supply function
$q^d = a + bp$
$q^s = c + dp^*$
where p is the price the consumer pays and p^* the price the producer receives for each item, and $p = p^* + t$ where t is the amount of tax on each item.

Substituting for p^* in the supply function, we can write
$q^s = c + d(p - t)$
or $q^s = c - dt + dp$.
Combining this with the demand function, we can calculate the equilibrium price as

$$p = \frac{a - c + dt}{d - b} \quad \text{(see section 4.9.4)}$$

which can also be written as

$$p = \frac{a - c}{d - b} + \frac{dt}{d - b}.$$

We are now in a position to use our differentiation to consider how this equilibrium price changes as the amount of tax imposed changes. Since we are interested in how p changes as t changes, we consider the above relation to be a function of t,

i.e. $p = \dfrac{a - c}{d - b} + \dfrac{dt}{d - b} = f(t).$

To find how p changes as t changes we consider how $f(t)$ changes as t changes, and this is given by the derivative $f'(t)$ which in this case is

$$f'(t) = \frac{d}{d-b}.$$

(Remember a, b, c and d are just some constant numbers, so $(a-c)/(d-b)$ and $d/(d-b)$ are just constant numbers.)

Now $d > 0$ and $b < 0$ (see section 4.9.3) and therefore $d/(d-b) > 0$. Therefore $f'(t) > 0$ and we conclude that the $f(t)$ function has a positive slope, i.e. as t increases $f(t)$ (or p) increases. So an increase in the amount of tax does increase the price the consumer pays.

The quantity demanded and supplied at the equilibrium price is

$$q^d = q^s = \frac{da - bc}{d-b} + \frac{bdt}{d-b} \text{ (see section 4.9.4)}$$

and a similar analysis reveals the effect of a change in t on the quantity supplied and demanded.

$$q^d = q^s = \frac{da - bc}{d-b} + \frac{bdt}{d-b} = g(t).$$

The derivative $g'(t) = \dfrac{bd}{d-b}$

and since $d > 0$ and $b < 0$ this expression is < 0.
Therefore $g'(t) < 0$
and we conclude that an increase in the amount of tax *reduces* the quantity sold. (See Fig. 4.17 for confirmation.)

5.6.6 Exercise

Given a simple market model
$q^d = a + bp$
$q^s = c + dp^*$.

Find the equilibrium price, and the corresponding quantity demanded and supplied, if the commodity has a *subsidy* of an amount s, for each unit supplied (i.e. $p = p^* - s$).

Using differentiation, consider the effects on the equilibrium price and quantity, of changes in the amount of subsidy.

Chapter 6

Maximisation and minimisation of functions

Economics has been described as the study of 'the optimum allocation of scarce resources'. The concept of optimisation is a recurrent one in economics, and in this chapter we consider differentiation as a mathematical technique for finding optimum values.

In section 6.6 we show how this technique can be applied to discover the level of output a firm should produce, if it is to maximise its profit. (We have already pointed out, in section 2.10.10, that one objective of a firm might be to maximise its profit.) We also show, in section 6.6.3, that if a firm is maximising its profit, then marginal revenue will be equal to marginal cost, a proposition which should be familiar to readers who have studied the theory of the firm. Finally, the relationship, between the marginal cost and average cost functions is investigated in section 6.6.5.

6.1 Introduction

In this chapter we shall see how differentiation can be used for finding maximum and minimum points for functions, a topic which is of great importance to economists (see, for example, section 6.6). Let us start with some definitions.

Definition
For a function $f(x)$, *a maximum point* is a value of x, call it x^*, such that $f(x^*) \geq f(x)$ for any other value of x.
N. B. We define the maximum point to be x^*, not $f(x^*)$.
Example
The function with the graph in Fig. 6.1 has a maximum point when $x = 1$, i.e. $x^* = 1$
and $f(x^*) = f(1) = 2$.

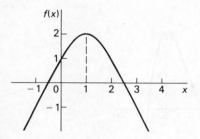

Fig. 6.1

Similarly, we define a minimum as:

Definition
For a function $f(x)$, *a minimum point* is a value of x, call it x^*, such that $f(x^*) \leq f(x)$ for any other value of x.

The points defined above are often referred to as *global maximum* and *global minimum*, because of the phrase 'for *any* other value of x'.

For the purposes of this chapter it will be more convenient to talk in terms of *local* maximum or minimum points.

Definition
For a function $f(x)$ a *local maximum point* is a value of x, call it x^*, such that $f(x^*) \geq f(x)$ for any other value of x *close to (or in the neighbourhood of) x^**.
Similarly:

Definition
For a function $f(x)$ a *local minimum point* is a value of x, call it x^*, such that $f(x^*) \leq f(x)$ for any other value of x, *close to (or in the neighbourhood of) x^**.

Example
For the function in Fig. 6.2, $x = -1$ is a *local maximum point*, $x = 2$ is a *local minimum point* and $x = 5$ is both a *local and a global* maximum point.

N. B. $x = 2$ is not a global minimum because although it corresponds to the smallest value of $f(x)$ in the neighbourhood of $x = 2$, it is not the smallest value for *any* other x, e.g. $x = -3$ has a smaller value of $f(x)$. This function does not have a global minimum.

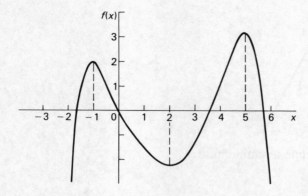

Fig. 6.2

N. B. A point which is a global maximum (or minimum) will also be a local maximum (or minimum) but a local maximum (or minimum) need not be a global maximum (or minimum).

In this chapter we shall derive a method for finding where the *local* maximum or minimum points of a function are. The method does not tell us directly whether such points are also global maximum or minimum points.

6.2 First order condition for a maximum or minimum point

6.2.1 *A maximum point*

Let us concentrate on a function $f(x)$ with one maximum point; its graph might look like Fig. 6.3.

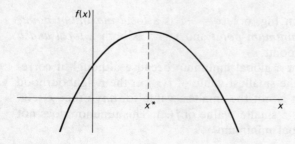

Fig. 6.3

140

We wish to find where the maximum point x^* is. We shall locate this point by observing what happens to the *slope* of $f(x)$ close to this point.

For *values of x less than x^*, the slope of $f(x)$ is positive.*

Example

At points A and B the slope is positive (although smaller at B than at A) (Fig. 6.4).

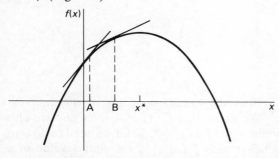

Fig. 6.4

For *values of x greater than x^*, the slope of $f(x)$ is negative.*

Example

At points C and D the slope is negative (although smaller (more negative) at D than at C) (Fig. 6.5).

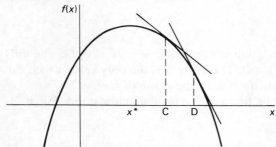

Fig. 6.5

So to the left of x^* the slope is positive, to the right it is negative and *at x^* the slope is zero* (Fig. 6.6).

We have then that *at a maximum point, the slope of $f(x)$ is zero*, i.e. $f'(x^*) = 0$.

This is known as the *first order condition for a maximum*. (We shall meet the second order condition later in this chapter.)

We can use this condition to locate the maximum point of a function.

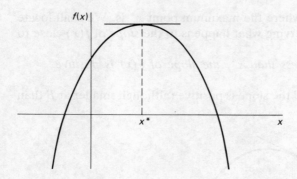

Fig. 6.6

Example

Suppose we wish to find the maximum point of the function $f(x) = 2 + 4x - x^2$. We know that $f'(x) = 4 - 2x$ and we also know that at a maximum point $f'(x) = 0$. So all we need to do is to find a value of x which makes $f'(x) = 0$, i.e. find a value of x which makes $4 - 2x = 0$. But this is precisely the problem we discussed in Chapter 4 on equations. 'Solve $4 - 2x = 0$' is in fact a linear equation which can be solved by manipulation:

$$4 - 2x = 0$$
$$-2x = -4$$
$$2x = 4$$
$$x = 2.$$

So $x = 2$ is the only value which makes $4 - 2x = 0$, i.e. the only value at which $f'(x) = 0$. Therefore it is the only value which can be a maximum point. If we draw the graph of $f(x) = 2 + 4x - x^2$ we do get confirmation that $x = 2$ is the maximum point (Fig. 6.7).

Summary

To find a maximum point of a function $f(x)$ we differentiate the function to get $f'(x)$, set $f'(x) = 0$ and solve the resulting equation.

Remark

We shall see in section 6.3 that the above procedure, of finding $f'(x)$, setting $f'(x) = 0$ and solving the equation, does not always guarantee that the resulting point is a maximum. We shall ignore this problem for the moment, but the reader is advised to read section 6.3 before using the technique.

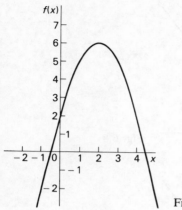

Fig. 6.7

Example
Find the maximum point of $f(x) = 1 - 2x - x^2$.
Now $f'(x) = -2 - 2x$
and $f'(x) = 0$ when $-2 - 2x = 0$
or when $-2x = 2$
or $2x = -2$
or $x = -1$.
Therefore $x = -1$ is the maximum point.

6.2.2 Exercises

Find the maximum points for the following functions:
1. $f(x) = 4 - 8x - x^2$
2. $f(x) = 2 + 3x - x^2$.

6.2.3 A minimum point

We shall now carry out a similar analysis to discover how to locate a minimum point and will find that we can use *exactly the same procedure as for a maximum*.
Consider a function $f(x)$ with a minimum point at x^* (Fig. 6.8).
At values of *x to the left of x^*, the slope of $f(x)$ is negative* (Fig. 6.9).
To the right of x^* the slope of $f(x)$ is positive (Fig. 6.10).
And *at the minimum point x^* the slope is again zero* (Fig. 6.11), i.e. $f'(x^*) = 0$.

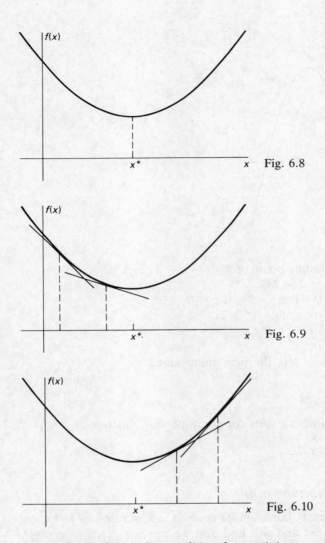

Fig. 6.8

Fig. 6.9

Fig. 6.10

This is the *first order condition for a minimum* and is of course identical to the first order condition for a maximum, and can be used in the same way to locate minimum points.

Remark
We shall see in section 6.3 that this procedure, for finding minimum points, does not guarantee that the point found is a

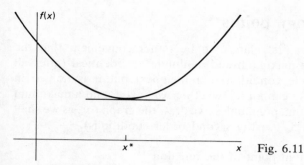

Fig. 6.11

minimum. We shall again ignore this problem for the present but the reader is advised to consult section 6.3 before using the technique.

Example

Find the minimum point of $f(x) = x^2 - 6x + 3$.

Now $f'(x) = 2x - 6$

and $f'(x) = 0$ when $2x - 6 = 0$,

i.e. when $2x = +6$

or $x = 3$.

So $x = 3$ is the minimum point, which is verified by looking at the graph of $f(x) = x^2 - 6x + 3$ (Fig. 6.12).

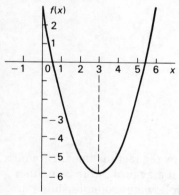

Fig. 6.12

6.2.4 Exercises

Find the minimum points for the following functions:
1. $f(x) = x^2 + 8x + 2$
2. $f(x) = x^2 - 3x + 1$.

6.3 Stationary points

It may seem, at first glance, to be rather convenient that the conditions for a maximum and minimum are identical (after all there is only one condition to remember) but it is not so in practice, since we cannot tell whether we have found a maximum point or a minimum point unless we draw the graph (or, as we shall see in section 6.5, use the second order conditions).

Example
Find the maximum point of the function
$f(x) = \frac{1}{3}x^3 - 2x^2 + 3x - 1$.
Now $f'(x) = x^2 - 4x + 3$
and $f'(x) = 0$ when $x^2 - 4x + 3 = 0$.
Here we have a *quadratic* equation which is solved by either factorisation or formula (see sections 4.3.2 or 4.3.3) to give two solutions
$\quad x = 1 \quad$ or $\quad x = 3$.
So there are two values of x at which $f'(x) = 0$. Which (if either) is the maximum? Only by drawing the graph of $f(x)$ can we decide that $x = 1$ is the (local) maximum (Fig. 6.13).

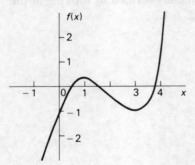

Fig. 6.13

The situation is further complicated by the fact that having found a value of x at which the slope is zero, it may turn out to be neither a maximum nor a minimum as the following example shows.

Example
Has the function $f(x) = x^3 - 3x^2 + 3x + 1$ got a maximum point? If we follow the procedure for finding the maximum point we get:
$f'(x) = 3x^2 - 6x + 3$
$f'(x) = 0$ when $3x^2 - 6x + 3 = 0$

which yields a quadratic equation, and when solved (either by factorisation or formula) yields just one solution $x = 1$.

So there is just one value of x which makes $f'(x) = 0$, namely $x = 1$. We must now ask whether this point is a maximum or a minimum. If we draw the graph of $f(x)$ we find that it is in fact neither (Fig. 6.14).

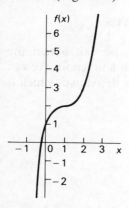

Fig. 6.14

The point is what is known as a point of *inflexion*, it is neither a maximum nor a minimum point (although the slope at the point is zero). If, then, we find a value of x at which $f'(x) = 0$ it could be that we have found a maximum or a minimum or even a point of inflexion. It is convenient to refer to all of these cases as stationary points.

Definition

A *stationary point* is a value of x at which $f'(x) = 0$.

N. B. A stationary point can be a maximum or a minimum or a point of inflexion. We shall discuss how to distinguish between the different types (by using second order conditions) in section 6.5. In section 6.4 we shall do some necessary groundwork.

6.3.1 Exercises

Find the stationary points for the following functions. Draw the graphs of the functions to determine the type of stationary point found.

1. $f(x) = x^2 + 4x - 1$
2. $f(x) = 2 - x - x^2$
3. $f(x) = x^3 - 3x + 1$ (this should have *two* stationary points)
4. $f(x) = x^3 + 6x^2 + 12x + 1$.

6.4 Second and higher derivatives

We have seen how, given a function $f(x)$, we can obtain the derivative $f'(x)$ and that the derivative is itself a function. We can differentiate this second function to obtain *its* derivative, which is denoted by $f''(x)$.

Example
If $\quad f(x) = 4x^3 - x^2 + 3x - 1$
then $\ f'(x) = 12x^2 - 2x + 3$
and differentiating again
$\quad f''(x) = 24x - 2$.

Definition
$f''(x)$ is known as the *second derivative* of $f(x)$, and gives the slope (or rate of change) of $f'(x)$.

Example
If $\quad f(x) = 1 - 3x$
then $\ f'(x) = -3$
and $\ f''(x) = 0$.

Example
If $\quad f(x) = x^5 - 4x^3 + 7x^2 - x + 2$
then $\ f'(x) = 5x^4 - 12x^2 + 14x - 1$
and $\ f''(x) = 20x^3 - 24x + 14$.

Example
If $\quad f(x) = \log_e(x^2)$

then $\ f'(x) = \dfrac{1}{x^2} \cdot 2x$ (using the function of a function rule),

i.e. $\ f'(x) = \dfrac{2}{x} = 2 \cdot x^{-1}$

and $\ f''(x) = -2x^{-2} = \dfrac{-2}{x^2}$.

Remark

An alternative notation for the second derivative is d^2y/dx^2.

Example

A function $f(x) = x^3 - 2x^2 + x - 5$ can be written
$$y = x^3 - 2x^2 + x - 5$$

and then $\quad \dfrac{dy}{dx} = 3x^2 - 4x + 1$

and $\quad \dfrac{d^2y}{dx^2} = 6x - 4.$

Remark

Since the second derivative is also a function, we could differentiate again to get $f'''(x)$, the third derivative of $f(x)$, and indeed repeated differentiation would yield higher order derivatives.

Remark

Not all functions can be differentiated, and those that do have first derivatives do not always have second derivatives. Similarly those that do have second derivatives do not always have third derivatives, etc.

6.4.1 Exercises

Find the second derivatives for the following functions:
1. $f(x) = x^4 - 3x^2 + 2$
2. $f(x) = 1 - x$
3. $f(x) = e^{x^2}$
4. $f(x) = \cos(x)$

6.5 Second order conditions for maximum and minimum points

6.5.1 A maximum

Consider a maximum point (Fig. 6.15).

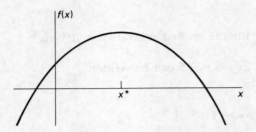

Fig. 6.15

We have already seen (in section 6.2.1) that for x values below x^* the slope of $f(x)$ (i.e. $f'(x)$) is positive, and for values of x greater than x^*, the slope ($f'(x)$) is negative, whilst *at* x^* the slope ($f'(x^*)$) is zero. We shall now analyse in more detail how $f'(x)$ changes. At points to the left of x^*, $f'(x)$ is positive, and as we move closer to x^* (i.e. increase x) the slope ($f'(x)$) decreases until at x^* it is zero. As x increases above x^*, the slope ($f'(x)$) is negative becoming more so as x continues to increase. In other words, as x increases $f'(x)$ decreases (from positive, through zero, to negative). Therefore the slope of $f'(x)$ is negative. Therefore $f''(x) < 0$. (Recall that $f''(x)$ gives the slope of $f'(x)$.)

We expect then, that *at a maximum point* x^*, $f''(x^*) < 0$. (Strictly speaking, it is possible to have a maximum when $f''(x) = 0$. An example is given later.)

This is known as the *second order condition for a maximum*. (Recall that the first order condition is that $f'(x^*) = 0$.)

Example

In section 6.2.1 we considered $f(x) = 2 + 4x - x^2$

which has $f'(x) = 4 - 2x$

and $f'(x) = 0$ when $x = 2$.

Therefore $f(x)$ has a stationary point when $x = 2$ and a graph of the function revealed this to be a maximum. We can now check the second order condition.

Since $f'(x) = 4 - 2x$

$f''(x) = -2$,

i.e. $f''(x)$ is negative for all values of x and in particular, at our stationary point $x = 2, f''(x) < 0$ which confirms the second order condition.

6.5.2 A minimum

Consider now a minimum point (Fig. 6.16).

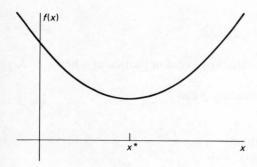

Fig. 6.16

As x increases from below x^*, the slope changes from being negative, to zero at x^*, to positive above x^*. Closer inspection reveals that as x increases the slope (i.e. $f'(x)$) starts negative, becomes less negative (i.e. increases) as x increases until the slope is zero, and then carries on increasing above x^*. In other words, as x increases $f'(x)$ increases (from negative, through zero, to positive). Therefore the slope of $f'(x)$ is positive. Therefore $f''(x) > 0$.

We expect, then, that *at a minimum point x^*, $f''(x^*) > 0$*. (Strictly speaking, it is possible to have a minimum when $f''(x) = 0$. An example is given later.) Comparison of this *second order condition for a minimum* with the second order condition for a maximum (which was that $f''(x^*) < 0$), reveals that while the first order condition is identical for both (namely that $f'(x^*) = 0$), the second order conditions are different. The difference in the second order condition can then be used to determine whether a stationary point is a maximum or a minimum.

Example

Consider the function
$f(x) = 4 - 6x + x^2$.
To find the stationary point
$f'(x) = -6 + 2x$.
At a stationary point $\qquad f'(x) = 0$.
So we want x such that $\quad -6 + 2x = 0$
$$2x = 6$$
$$x = 3.$$
There is a stationary point when $x = 3$.
To determine whether it is a maximum or minimum consider the

second derivative:
$$f'(x) = -6 + 2x.$$
Therefore $f''(x) = +2$,
i.e. the $f''(x) > 0$ for all values of x and in particular when $x = 3$, $f''(x) > 0$.
Therefore $x = 3$ is a minimum point.

Example
$f(x) = \frac{1}{3}x^3 - 2x^2 + 3x - 1$.
We saw in section 6.3 that
$$f'(x) = x^2 - 4x + 3$$
and that $f'(x) = 0$ when $x^2 - 4x + 3 = 0$.
This quadratic equation had two solutions:
  $x = 1$ and $x = 3$.
There are *two* stationary points $x = 1$ and $x = 3$.
To determine what kind of stationary points they are, we examine the second derivative.
Since $f'(x) = x^2 - 4x + 3$
then $f''(x) = 2x - 4$.
When $x = 1$ $f''(x) = 2 \cdot 1 - 4 = -2$,
i.e. $f''(x) < 0$, therefore $x = 1$ is a maximum.
When $x = 3$ $f''(x) = 2 \cdot 3 - 4 = 2$,
i.e. $f''(x) > 0$, therefore $x = 3$ is a minimum.
Examination of the graph of $f(x)$ confirms this (Fig. 6.17).

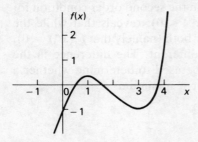

Fig. 6.17

N. B. We must first find the stationary points, using the first order condition, and then apply the second order condition to each stationary point to determine whether it is a maximum or minimum.

We have stated that if at some value x^*,
  $f'(x^*) = 0$ and $f''(x^*) < 0$, we have a maximum
or if $f'(x^*) = 0$ and $f''(x^*) > 0$, we have a minimum.

The discerning reader will have observed that we have made no mention of what happens if $f'(x^*) = 0$ and $f''(x^*) = 0$. Neither have we discussed a second order condition for a point of inflexion. It would be rather nice if we could state that 'if $f'(x^*) = 0$ and $f''(x^*) = 0$ then we have a point of inflexion'. Unfortunately this is *not* so as the following example illustrates.

Example

$f(x) = x^4$

To find the stationary point consider $f'(x)$.

$f'(x) = 4x^3$
$f'(x) = 0$ when $4x^3 = 0$,
i.e. when $x = 0$.

So we have only one stationary point, $x = 0$. To determine what type of stationary point, consider $f''(x)$.

Since $f'(x) = 4x^3$
then $f''(x) = 12x^2$.

At the stationary point $x = 0$, $f''(x) = 12 \cdot 0 = 0$.

So we have a point $x = 0$, at which $f'(x) = 0$ and $f''(x) = 0$ but the point is *not* a point of inflexion as the graph shows (Fig. 6.18).

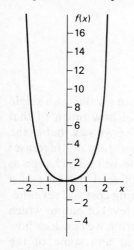

Fig. 6.18

The above example demonstrates that if, at a point x^*, $f'(x^*) = 0$ and $f''(x^*) = 0$ we cannot conclude that the point is a point of inflexion although it *may* be.

If at a stationary point x^*, $f''(x^*) = 0$, consideration of *higher order derivatives* is necessary before a conclusion about the type of stationary point can be reached. We shall not concern ourselves here with higher order tests since they are seldom required in economics. (The interested reader can consult, for example, R. G. D. Allen's *Mathematical Analysis for Economists*.)

6.5.3 Exercises

Find the stationary points for the following functions and use the second order conditions to determine the type of stationary point.

1. $f(x) = x^2 - 6x + 1$
2. $f(x) = 2 + x - x^2$
3. $f(x) = \dfrac{x^3}{3} + 2x^2 + 3x - 1$
4. $f(x) = 1 - 2x^4$
5. $f(x) = \dfrac{x^3}{3} + x^2 + x - 1$.

Draw the graphs of the last two functions.

6.6 Applications to economics

6.6.1 Profit maximisation

In section 2.10 we introduced the idea of a firm producing a single commodity, being faced with the decision of how much of that commodity it should produce. In section 2.10.6 we saw that as the firm increases output so its costs will change (usually increase) and in section 2.10.8 we saw that its revenue will also change as output changes. In section 2.10.10 we introduced the idea of a profit function to take into account this changing cost and revenue, and suggested that a firm might produce that level of output which made its profit as large as possible. In this section we consider how the firm can determine this level of output and some of the consequences of its choosing 'the output which maximises profit'.

Example

A firm is faced with the following cost and demand functions:
$$\text{Cost} = x^3 - 9x^2 + 30x + 10.$$
Demand function, $\quad p = 70 - 8x$.

The total revenue function is obtained from the demand function (see section 2.10.8) as:

revenue $= p \cdot x = (70 - 8x) \cdot x$
$= 70x - 8x^2$

and consequently the profit function will be (see section 2.10.10)

profit = revenue − cost
$= (70x - 8x^2) - (x^3 - 9x^2 + 30x + 10)$
$= 70x - 8x^2 - x^3 + 9x^2 - 30x - 10$

profit $= 40x + x^2 - x^3 - 10$.

To find the maximum of *any* function we first of all find the stationary points using the method of section 6.2.

$$f(x) = 40x + x^2 - x^3 - 10.$$

Therefore $f'(x) = 40 + 2x - 3x^2$.

Therefore $f'(x) = 0$ when $40 + 2x - 3x^2 = 0$.

We have a quadratic equation to solve, therefore we shall use the formula given in section 4.3.3 (although it does in fact factorise). To use the formula we must re-write the equation in the correct order.

So $40 + 2x - 3x^2 = 0$ becomes
$-3x^2 + 2x + 40 = 0$

and the solutions are given by

$$x = \frac{-2 \pm \sqrt{4 - 4 \cdot (-3) \cdot 40}}{2 \cdot -3} \quad \text{(see section 4.3.3)}$$

$$= \frac{-2 \pm \sqrt{4 + 480}}{-6}$$

$$= \frac{-2 \pm \sqrt{484}}{-6}$$

$$= \frac{-2 \pm 22}{-6}.$$

So the two solutions are

$$x = \frac{-2 - 22}{-6} = \frac{-24}{-6} = +4$$

or $\quad x = \dfrac{-2 + 22}{-6} = \dfrac{+20}{-6} = -3\tfrac{1}{3}$.

There are then, two stationary points of the profit function, $x = 4$ and $x = -3\frac{1}{3}$ (though an output of $-3\frac{1}{3}$ units is clearly implausible).

We must now use the second order conditions, on the profit function, to determine whether either is a maximum point (as in section 6.5.2).

The profit function was
$$f(x) = 40x + x^2 - x^3 - 10$$
and $$f'(x) = 40 + 2x - 3x^2.$$
Therefore $f''(x) = 2 - 6x.$
When $x = 4$, $f''(x) = 2 - 6 \cdot 4 = -22,$
i.e. $f''(x) < 0$, therefore $x = 4$ *is a maximum*.
When $x = -3\frac{1}{3}$ $f''(x) = 2 - 6 \cdot (-3\frac{1}{3}) = +22,$
i.e. $f''(x) > 0$, therefore $x = -3\frac{1}{3}$ is a minimum.

We have discovered that the value of x which maximises $f(x)$ is $x = 4$, i.e. the output the firm must produce in order to maximise profit is 4 units. At this level of output the profit, given by the profit function, will be

Profit $= 40 \cdot 4 + 4^2 - 4^3 - 10$
$= 102.$

N. B. In the example, to find the stationary point, we had to solve a *quadratic* equation. In general the type of equation which occurs will depend only on the types of functions which specify the demand and cost functions. Exercise 6.6.2 provides an example where the equation is not quadratic.

Remark

The above procedure for finding the level of output which will produce maximum profit can only be used if it is possible to obtain a profit function which can be differentiated.

Remark

The fixed cost plays no part in the determination of the profit maximising output. Since the fixed cost is a constant (i.e. independent of x) in the cost function, it is also a constant in the profit function, and therefore disappears on differentiation. Consequently whatever the size of the fixed cost, the output which maximises profit will be the same (although the profit made at this output will differ).

Example
In the previous example the firm's fixed costs were 10. If another firm has exactly the same demand and *variable* cost functions but has a fixed cost of 210, it will also produce 4 units of output to maximise its profit. So its output is the same but its profit will not be, since its profit (as given by its profit function) will be -98, i.e. it will make a loss! This is not inconsistent with the concept of profit maximisation, all we are saying is that the best this firm can do is make a loss of only 98. Any other output would produce an even bigger loss.

6.6.2 Exercise

A firm is faced with the following cost function and demand function:
$p = 100 - \frac{1}{2}x$
$\text{cost} = \frac{1}{2}x^2 + 10x + 50.$
Find the total revenue function and the profit function. Hence find the level of output which will maximise profit, and the profit the firm will make if it produces this output.

6.6.3 Relation between marginal cost and marginal revenue

In section 5.6 we defined the marginal cost and marginal revenue functions. In this section we show that if a firm produces the output which maximises profit, then at this output the marginal revenue is equal to the marginal cost.

We shall denote the total cost and total revenue functions by f and g respectively:
$\text{cost} = f(x)$
$\text{revenue} = g(x)$, where x is output.
Then $\text{profit} = \text{revenue} - \text{cost}$
$= g(x) - f(x),$
which is, of course, also a function of output x.

To find the value of x which produces maximum profit we use the first order condition of section 6.2.1, i.e. we differentiate the profit function and set the derivative equal to zero. Differentiating the profit function (using the rule in section 5.5.3) we get

$g'(x) - f'(x)$, and setting the derivative equal to zero gives $g'(x) - f'(x) = 0$.

So the value of x which maximises profit is one which satisfies the equation $g'(x) - f'(x) = 0$, or the equation $g'(x) = f'(x)$ (using the rule (a) in section 4.2.2 for manipulating equations). But $g'(x)$ is the marginal revenue function and $f'(x)$ is the marginal cost function.

In other words the value of x which maximises profit is one such that the marginal cost is equal to the marginal revenue.

Example

In the first example of section 6.6.1 the $f(x)$ and $g(x)$ functions were given by:

$\quad$ cost $= x^3 - 9x^2 + 30x + 10$
revenue $= 70x - 8x^2$.

Consequently:

$\quad\quad$ marginal cost $= 3x^2 - 18x + 30$
and $\quad$ marginal revenue $= 70 - 16x$.

The profit was maximised at $x = 4$ (see section 6.6.1). When $x = 4$ the marginal cost $= 3 \cdot 4^2 - 18 \cdot 4 + 30 = 6$ and when $x = 4$ the marginal revenue $= 70 - 16 \cdot 4 = 6$. Therefore marginal cost = marginal revenue.

Remark

This property provides an alternative method of determining the level of output which maximises profit, without actually working out the profit function as we did in section 6.6.1. We merely find the marginal revenue function, find the marginal cost function, equate the two and solve the resulting equation to find x. In the example above, equating marginal cost and marginal revenue produces

$\quad 3x^2 - 18x + 30 = 70 - 16x$
or $\quad 3x^2 - 2x - 40 = 0$.

Solving this equation gives two solutions $x = 4$, $x = -3\frac{1}{3}$.

The drawback to this method is that we shall have to either use the second order conditions of the previous method to determine which of these is the maximum (the result that marginal cost = marginal revenue is based only on the first order condition for a maximum which is of course, identical to the first order condition for a minimum) *or* use the condition that at a *maximum* we expect $g''(x) - f''(x) < 0$, i.e. $g''(x) < f''(x)$: i.e. the slope of the marginal revenue function is less than the slope of the marginal cost function.

6.6.4 Exercise

For the example of exercise 6.6.2 show that marginal cost is equal to marginal revenue at the output which maximises profit (i.e. at the output $x = 45$).

6.6.5 The relation between the marginal and average cost curves

In this section we shall demonstrate another well-known property of microeconomics, namely, that the marginal cost equals the average cost when the latter is at its minimum value.

Suppose the total cost function is $f(x)$.
i.e. $\qquad \text{cost} = f(x)$.

Then $\quad$ average cost $= \dfrac{f(x)}{x}$,

which is another function of x.

To find the minimum of this average cost function, we use the method of section 6.2.3, i.e. differentiate the function, set the derivative equal to zero and solve the equation.

The derivative of the average cost function (using rule 5.5.6) is

$$\frac{f'(x) \cdot x - 1 \cdot f(x)}{x^2}$$

and setting this to zero gives the equation

$$\frac{f'(x) \cdot x - f(x)}{x^2} = 0,$$

i.e. average cost is minimised at a value of x which satisfies this equation.

This equation is certainly not of a type we have met before, but we can do a certain amount of manipulation on it using the rules in section 4.2.2.

Multiply both sides by x^2
$f'(x) \cdot x - f(x) = 0$.

Add $f(x)$ to both sides
$$f'(x) \cdot x = f(x).$$
Divide both sides by x (assuming that x is not zero)
$$f'(x) = \frac{f(x)}{x}.$$

So average cost is minimised at a value of x which satisfies this equation. But this equation states that marginal cost ($f'(x)$) is equal to average cost ($f(x)/x$). So we have shown that at the value of x which minimises average cost, the average cost is equal to the marginal cost.

Example

If the total cost function is
$$\text{cost} = x^3 - 4x^2 + 10x$$
then marginal cost $= 3x^2 - 8x + 10$

and average cost $= \dfrac{x^3 - 4x^2 + 10x}{x} = x^2 - 4x + 10.$

Average cost is minimised when the derivative of the average cost function equals zero, i.e. when
$$2x - 4 = 0$$
giving $\qquad x = 2.$

(The second derivative verifies that this is indeed a minimum.)
At this value average cost $= 2^2 - 4 \cdot 2 + 10 = 6$
and marginal cost $= 6$.

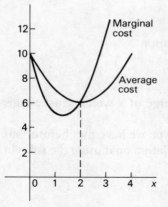

Fig. 6.19

So at the value of output which minimises average cost, average cost and marginal cost are equal.

The graphs of the average and marginal cost functions are given in Fig. 6.19.

6.6.6 Exercise

For the following total cost function find the average and marginal cost functions:
$$\text{cost} = 2x^3 - 20x^2 + 100x.$$
Find the value of x which minimises average cost. Find the average and marginal cost at this value of x.

Chapter 7

Integration

In Chapter 5 we discovered how, given many functions $g(x)$, we could find their derivatives $g'(x)$, and we used the idea to obtain marginal revenue and marginal cost functions for a firm, from total revenue and total cost functions. In this chapter we define *integration* to be the reverse of differentiation and hence we shall be able to obtain the total revenue and total cost functions from the appropriate marginal functions. In addition we shall see how integration can be used in considering the relationship between investment and capital accumulation.

7.1 Introduction

Given almost any function $g(x)$ we can find its derivative $g'(x)$ by the process of differentiation.
Example
If $g(x) = 3x^2 + 2x - 4$ then $g'(x) = 6x + 2$.

In this chapter we reverse the process, i.e. we shall be given $g'(x)$ and required to find $g(x)$. This reverse process is called *integration*. For example, we might be given that $g'(x) = 6x + 2$ and asked to find $g(x)$.

Notation
Instead of writing 'If $g'(x) = 6x + 2$ what is $g(x)$?', we write

$$\int (6x + 2) dx$$

where '$\int$' is known as the *integral* sign
and we say we *integrate* $6x + 2$.

(The dx bit refers to the fact that we are integrating 'with respect to x'. The significance of this will become more apparent in section 7.4.2.)

Let us have a formal definition of the integration process.

Definition
The integral of a function $f(x)$, written $\int f(x)\,dx$, is another function $F(x)$ such that the derivative of $F(x)$ is $f(x)$.

i.e. $$\int f(x)\,dx = F(x)$$

where $F(x)$ is such that
$F'(x) = f(x)$.
We say we *integrate* $f(x)$ to get $F(x)$.

Example

$$\int 3x^2\,dx = x^3$$

because if we differentiate $F(x) = x^3$ we get $F'(x) = 3x^2$.

Example

$$\int (3x^2 - 2x + 1)\,dx = x^3 - x^2 + x$$

because if we differentiate $F(x) = x^3 - x^2 + x$ we get back to $3x^2 - 2x + 1$.

Example

$$\int \cos(x)\,dx = \sin(x)$$

because if we differentiate $F(x) = \sin(x)$ we get back to $\cos(x)$.

N. B. So far we have not said how, given $f(x)$ we would actually find $F(x)$. We shall discuss this in the next section.

7.2 Rules of integration

Since integration is the reverse of differentiation, rules for integrating functions can be derived by reversing the rules of differentiation given in section 5.5 and the reader should compare the rules presented here with those.

7.2.1 The power rule

$$\int ax^n dx = \frac{ax^{n+1}}{n+1}$$

where a and n are *any* two numbers *except* that n cannot be equal to -1. (We say we 'raise the power by one, and divide by the new power'.)

Example

$$\int 3x^2 dx = \frac{3x^{2+1}}{2+1} \text{ (here } a = 3, n = 2)$$

$$= \frac{3x^3}{3} = x^3.$$

(We can easily check that if we differentiate $F(x) = x^3$ we do get back to $3x^2$.)

Example

$$\int 4x^5 dx = \frac{4x^6}{6} \text{ (here } a = 4, n = 5)$$

$$= \frac{2x^6}{3}.$$

Example

$$\int \frac{1}{x^2} dx = \int x^{-2} dx \text{ (here } a = 1, n = -2)$$

$$= \frac{x^{-1}}{-1} = -\frac{1}{x} \left(\text{since } x^{-1} = \frac{1}{x} \right).$$

Remark

The rule will not work if $n = -1$. The reason is that if we tried to apply the rule we should raise the power by one, to zero, and then be required to divide by zero, which is an operation not allowed in mathematics.

Special case

$$\int x\,dx = \frac{x^2}{2} \text{ because we can write}$$

$$\int x\,dx = \int 1 \cdot x^1\,dx$$

$$= 1 \cdot \frac{x^2}{2} = \frac{x^2}{2}.$$

Special case

$$\int a\,dx = ax \text{ because we can write}$$

$$\int a\,dx = \int ax^0\,dx \text{ (recall that } x^0 = 1 \text{ for any value of } x\text{)}$$

$$= \frac{ax^1}{1} = ax.$$

7.2.2 Integration of sums

$$\int (g(x) + h(x))\,dx = \int g(x)\,dx + \int h(x)\,dx,$$

i.e. we can 'integrate term by term'.

Example

$$\int (3x^2 + 2x)\,dx = \int 3x^2\,dx + \int 2x\,dx$$

$$= x^3 + x^2 \text{ (using the power rule)}.$$

Example

$$\int (x + 3)\,dx = \int x\,dx + \int 3\,dx$$

$$= \frac{x^2}{2} + 3x.$$

The rule extends naturally to integrating a sum of three (or more) terms.

Example

$$\int (3x^2 + 2x + 1)\,dx = \int 3x^2\,dx + \int 2x\,dx + \int 1\,dx$$
$$= x^3 + x^2 + x.$$

7.2.3 Integration of differences

$$\int (g(x) - h(x))\,dx = \int g(x)\,dx - \int h(x)\,dx.$$

Example

$$\int (x^3 - 2x)\,dx = \int x^3\,dx - \int 2x\,dx$$
$$= \frac{x^4}{4} - x^2.$$

As in section 5.5.3 we can extend the rule and combine it with the previous rule.

Example

$$\int (x^3 - x^2 + 2x - 1)\,dx = \int x^3\,dx - \int x^2\,dx + \int 2x\,dx - \int 1\,dx$$
$$= \frac{x^4}{4} - \frac{x^3}{3} + x^2 - x.$$

The three rules so far given are straightforward applications of the first three rules of differentiation, to integration, but as we shall see, the rules in sections 5.5.5. to 5.5.8 are more difficult to apply to integration. We shall therefore leave these until later in the chapter and discuss first the counterparts of the rules for functions given in section 5.5.9.

7.2.4 Integration of special functions ($\sin(x)$, $\cos(x)$, $\log_e x$, e^x)

These rules of integration follow directly from the corresponding rules of differentiation.

(a) $\int e^x dx = e^x$

(b) $\int \frac{1}{x} dx = \log_e x$

N. B. This rule takes care of the exception to the power rule, i.e. the integration of x^n when $n = -1$. It is also only valid for $x > 0$.

(c) $\int \sin(x) dx = -\cos(x) dx$

(d) $\int \cos(x) dx = \sin(x) dx$.

7.3 The arbitrary constant of integration

We have defined integration as the reverse of differentiation, i.e.

$\int f(x) dx = F(x)$ where $F(x)$ is such that

$F'(x) = f(x)$.

So, for example,

$\int (2x + 1) dx = x^2 + x$

because if we differentiate $F(x) = x^2 + x$ we do get back to $2x + 1$. However, the discerning reader may well ask him/herself why we cannot equally have

$\int (2x + 1) dx = x^2 + x + 2$

since when we differentiate $F(x) = x^2 + x + 2$ we also get back to $2x + 1$.

The answer is that we *can* have

$\int (2x + 1) dx = x^2 + x + 2$

or even $\int (2x + 1)\,dx = x^2 + x + 10$

or indeed $\int (2x + 1)\,dx = x^2 + x +$ any constant.

The reason is, of course, that on differentiation whatever constant we include will disappear.
So we write $\int (2x + 1)\,dx = x^2 + x + c$, where c stands for any constant value and is known as the *arbitrary constant of integration*.

Example

Find $\int (4x - 1)\,dx$.

Using rules in sections 7.2.3 and 7.2.1 we have

$$\int (4x - 1)\,dx = \frac{4x^2}{2} - x = 2x^2 - x$$

and allowing for the arbitrary constant

$$\int (4x - 1)\,dx = 2x^2 - x + c.$$

$f(x) = 2x^2 - x + c$ gives rise to a whole series of functions, one for each choice of c. The graphs of some of these functions are given in Fig. 7.1. Although each different choice for c gives rise to a different function (and hence a different graph), every one of these functions has one thing in common – namely, that for *any* value of x the slope at that point is the same for *all* the functions. For example, when $x = 1$, the slope is 3 for all of them. In integration problems all we know is the slope of the function, i.e. that the slope is given by $4x - 1$, and consequently the function we are interested in could be any function of the type $f(x) = 2x^2 - x + c$.

If we know just *one* ordered pair for the function, then we can establish precisely which function is relevant. For example, if we know that the graph passes through the point $(0, 4)$, then c must be equal to 4.

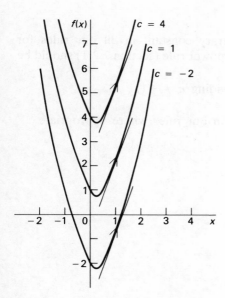

Fig. 7.1

Example

Find $\int (3x^2 - 2x + 1)\,dx$.

$$\int (3x^2 - 2x + 1)\,dx = \frac{3x^3}{3} - \frac{2x^2}{2} + x + c$$
$$= x^3 - x^2 + x + c.$$

If in addition we know that $F(x) = 4$ when $x = 1$, we can find the value of c, as follows.
$$F(x) = x^3 - x^2 + x + c.$$
When $x = 1$, then $F(x) = 1^3 - 1^2 + 1 + c$
$$= 1 + c.$$
But $F(x)$ must be equal to 4 when $x = 1$.
Therefore $1 + c = 4$.
Therefore $c = 3$

and $\int (3x^2 - 2x + 1)\,dx = x^3 - x^2 + x + 3$.

Remark

We should include the arbitrary constant in all our rules for integration. For example, the power rule (section 7.2.1) should be

$$\int ax^n dx = \frac{ax^{n+1}}{n+1} + c \quad \text{providing } n \neq -1.$$

In future we shall include c in our rules where appropriate.

7.3.1 Exercises

Find the following:

1. $\int (1 - x) dx$

2. $\int (x^2 - 3x + 2) dx$

3. $\int \frac{1}{x^3} dx$

4. $\int 4 dx$

5. $\int \left(x^3 - 3x^2 + 4x - 7 + \frac{1}{x^2} \right) dx.$

7.4 More difficult rules of integration

7.4.1 Integration by parts

This rule is based on differentiation of a product (section 5.5.5), but as we shall see it is not as easy to operate.

$$\int g(x) \cdot h(x) dx = g(x) \cdot k(x) - \int k(x) \cdot g'(x) dx$$

where $k(x) = \int h(x) dx.$

Example

For $$\int x \cdot \cos(x) \, dx$$

we take $g(x) = x$ and $h(x) = \cos(x)$.

Then $g'(x) = 1$ and $k(x) = \int h(x) \, dx$

$$= \int \cos(x) \, dx$$

$$= \sin(x).$$

Using the integration by parts rule

$$\int g(x) \cdot h(x) \, dx = g(x) \cdot k(x) - \int k(x) \cdot g'(x) \, dx$$

we get $\int x \cdot \cos(x) \, dx = x \cdot \sin(x) - \int \sin(x) \cdot 1 \, dx$

$$= x \cdot \sin(x) - \int \sin(x) \, dx$$

$$= x \cdot \sin(x) - (-\cos(x)) \text{ (using the rule in section 7.2.4).}$$

Therefore $\int x \cdot \cos(x) \, dx = x \cdot \sin(x) + \cos(x) + c.$

N. B. The arbitrary constant has been included at the final stage of integration.

Special case

$$\int c \cdot h(x) \, dx = c \cdot \int h(x) \, dx$$

for any function $h(x)$ and *any* constant c.

This can be demonstrated using the integration by parts rule with $g(x) = c$, or directly from the special case of the products rule of differentiation that if $f(x) = c \cdot g(x)$ then $f'(x) = c \cdot g'(x)$.

In particular, taking $c = -1$, we have

$$\int -h(x)\,dx = -\int h(x)\,dx.$$

Remark
Inspection of the integration by parts rule shows that the rule does not directly calculate the integral, it merely replaces the integral $\int g(x) \cdot h(x)\,dx$ by another $\int k(x) \cdot g'(x)\,dx$. The rule will only be of practical use if this second integral is easier to handle than the first. (Except in very exceptional cases – see the example below on $\int \sin(x)\cos(x)\,dx$.)

Remark
The choice of which of the two functions in the integral to take as $g(x)$ can be crucial (which is not the case in differentiation of a product) as the following example shows.

Example

$$\int \cos(x)\, x\,dx.$$

This is the same function as integrated in the previous example, however, this time we shall take $g(x) = \cos(x)$ and $h(x) = x$.

Now $g'(x) = -\sin(x)$ and $k(x) = \displaystyle\int h(x)\,dx = \int x\,dx = \frac{x^2}{2}$.

Using the rule

$$\int g(x) \cdot h(x)\,dx = g(x) \cdot k(x) - \int k(x) \cdot g'(x)\,dx$$

we get
$$\int \cos(x) \cdot x\,dx = \cos(x) \cdot \frac{x^2}{2} - \int \frac{x^2}{2} \cdot -\sin(x)\,dx$$
$$= \cos(x) \cdot \frac{x^2}{2} + \frac{1}{2}\int x^2 \cdot \sin(x)\,dx$$

(using the special case of section 7.4.1).

We have replaced $\int \cos(x) \cdot x\,dx$ with a more difficult integral $\int x^2 \cdot \sin(x)\,dx$ and clearly the integration by parts rule is not helpful in this case.

Remark
Sometimes repeated use of the rule may be necessary as the following example shows.
Example

To find $\int x^2 \sin(x)\,dx$

we take $g(x) = x^2$ and $h(x) = \sin(x)$.

Then $g'(x) = 2x$ and $k(x) = \int h(x)\,dx = \int \sin(x)\,dx$
$$= -\cos(x).$$

Then using the integration by parts rule

$$\int x^2 \sin(x)\,dx = x^2 \cdot -\cos(x) - \int -\cos(x) \cdot 2x\,dx$$

$$= -x^2 \cos(x) + 2\int \cos(x) \cdot x\,dx$$

(using the special case of section 7.4.1).
We must use the integration by parts rule again to find $\int \cos(x) \cdot x\,dx$, which in our first example in this section we showed to be equal to $x \sin(x) + \cos(x)$, and so

$$\int x^2 \sin(x)\,dx = -x^2 \cos(x) + 2(x \sin(x) + \cos(x)) + c$$
$$= -x^2 \cos(x) + 2x \sin(x) + 2 \cos(x) + c.$$

Remark
Sometimes, although the integration by parts rule does not produce an easier second integral, it can still prove beneficial as the following example shows.
Example

To find $\int \sin(x) \cdot \cos(x)\,dx$

take $g(x) = \sin(x)$ and $h(x) = \cos(x)$.

Then $g'(x) = \cos(x)$ and $k(x) = \int h(x)dx$

$$= \int \cos(x)dx = \sin(x).$$

Then, using integration by parts

$$\int \sin(x) \cdot \cos(x)dx = \sin(x) \cdot \sin(x) - \int \sin(x) \cdot \cos(x)dx.$$

This time we have replaced $\int \sin(x)\cos(x)dx$ by itself! However, all is not lost.

Let $I = \int \sin(x) \cos(x)dx$.

Then $I = \sin(x) \cdot \sin(x) - I$
or (adding I to both sides)
$\quad 2I = (\sin(x))^2$
or $\quad I = \frac{1}{2}(\sin(x))^2$,

i.e. $\int \sin(x) \cdot \cos(x)dx = \frac{1}{2}(\sin(x))^2$

or, if we include the arbitrary constant, $\int \sin(x) \cdot \cos(x)dx = \frac{1}{2}(\sin(x))^2 + c$.

7.4.2 The substitution rule

This rule is equivalent to the function of a function rule in section 5.5.8. (There is no rule in integration equivalent to 5.5.6, the quotient rule of differentiation.)

$$\int g(h(x)) \cdot h'(x)dx = \int g(u)du \text{ where } u = h(x).$$

The idea behind the rule is to replace the integral of a function in terms of x by the integral of a simpler function in terms of u where u is itself a function of x. Some examples should help to clarify the rule.

Example

To find $\int (2x^2 - 1)^3 \cdot 4x\,dx$.

Taking $h(x) = 2x^2 - 1$ and $g(x) = x^3$ this can be written as

$$\int g(h(x))h'(x)dx.$$

Hence, using our rule,

$$\int g(h(x)) \cdot h'(x)dx = \int g(u)du \text{ where } u = h(x) = 2x^2 - 1$$

we have

$$\int (2x^2 - 1)^3 \cdot 4x\,dx = \int u^3 du$$

$$= \frac{u^4}{4} + c \text{ (using the power rule)}$$

$$= \frac{(2x^2 - 1)^4}{4} + c \text{ (replacing } u \text{ by } h(x)).$$

Differentiation of $f(x) = \dfrac{(2x^2 - 1)^4}{4} + c$

does indeed give us $(2x^2 - 1)^3 \cdot 4x$ (using the function of a function rule).

Example

To find $\int 2x \cdot \cos(x^2 + 1)\,dx$.

Taking $h(x) = x^2 + 1$ and $g(x) = \cos(x)$ we can write the integral as

$$\int g(h(x)) \cdot h'(x)\,dx.$$

Using the rule,

$$\int g(h(x)) \cdot h'(x)dx = \int g(u)du \text{ where } u = h(x) = x^2 + 1$$

we have

$$\int 2x \cdot \cos(x^2 + 1)\,dx = \int \cos(u)\,du$$

$$= \sin(u) + c$$
(using the rule in section 7.2.4)
$$= \sin(x^2 + 1) + c.$$

Example

To find $\int \dfrac{-3x^2\,dx}{1 - x^3}$.

Taking $h(x) = 1 - x^3$ and $g(x) = 1/x$ we can write the integral as

$$\int g(h(x)) \cdot h'(x)\,dx.$$

Using the rule

$$\int g(h(x)) \cdot h'(x)\,dx = \int g(u)\,du \text{ where } u = h(x) = 1 - x^3$$

we have that

$$\int \frac{-3x^2\,dx}{1 - x^3} = \int \frac{1}{u}\,du$$

$$= \log_e u + c \text{ (using the rule in section 7.2.4)}$$
$$= \log_e(1 - x^3) + c.$$

N. B. The substitution rule will only work in simplifying the integral, in particular cases. If the original integral cannot be written as $\int g(h(x)) \cdot h'(x)\,dx$ then the rule cannot be used. For example, we could use the rule on $\int 2x \cdot \cos(x^2 + 1)\,dx$ (Example above), but could not have used it on $\int 2x^2 \cdot \cos(x^2 + 1)\,dx$.

7.4.3 Exercises

Find the following:

1. $\int x \cdot \sin(x)\,dx$

2. $\int x \cdot e^x dx$

3. $\int x^2 \cdot e^x dx$

4. $\int (x^3 + 2)^2 \cdot 3x^2 dx$

5. $\int \dfrac{2x + 1}{x^2 + x + 1} dx$

6. $\int 2x \cdot e^{x^2} dx$.

7.5 Definite integrals

The integrals we have discussed so far are sometimes referred to as *indefinite integrals*. An indefinite integral is a function. (Recall our definition in section 7.1.) In this section we introduce *definite integrals* which, as we shall see, are numbers, and in section 7.6 we shall see that these numbers are related to areas.

Let us introduce the idea of a definite integral using an example.
Example
Consider a function $f(x) = 2x + 1$, and the (indefinite) integral

$$\int (2x + 1) dx = x^2 + x + c = F(x).$$

Select *two* values for x. Say we choose $x = 1$ and $x = 3$. Substitute each in turn into the integral $F(x)$.
   When $x = 1$ $F(x) = 1^2 + 1 + c = 2 + c$
and   when $x = 3$ $F(x) = 3^2 + 3 + c = 12 + c$.
Finally, subtract the value of $F(x)$ associated with the smaller x value from the value of $F(x)$ associated with the larger x value, to give, in our example
   $(12 + c) - (2 + c) = 10$.
So we end up with a number. This number is known as the definite integral.

N. B. For a definite integral we need a function $f(x)$ and *two* values of x.

Definition
Given a function $f(x)$ and two values of x, $x = a$ and $x = b$ (with $a < b$), the *definite integral of $f(x)$ between the limits a and b* is defined as $F(b) - F(a)$ where $F(x)$ is the indefinite integral of $f(x)$.

Notation
The definite integral is written as $\int_a^b f(x)\,dx$.

Example

Find $\displaystyle\int_1^2 (x^2 + 2)\,dx$.

We first find the indefinite integral $F(x)$.

$$\int (x^2 + 2)\,dx = \frac{x^3}{3} + 2x + c = F(x).$$

Then evaluate $F(x)$ at 1 and 2 and subtract the former from the latter.

$F(1) = \tfrac{1}{3} + 2 + c = 2\tfrac{1}{3} + c$
$F(2) = \tfrac{8}{3} + 4 + c = 6\tfrac{2}{3} + c.$

Then $\displaystyle\int_1^2 (x^2 + 2)\,dx = F(2) - F(1)$

$$= (6\tfrac{2}{3} + c) - (2\tfrac{1}{3} + c)$$
$$= 6\tfrac{2}{3} + c - 2\tfrac{1}{3} - c$$
$$= 4\tfrac{1}{3}.$$

N. B. Notice how the arbitrary constant cancels out.

Notation
For the above example we would conventionally calculate the definite integral using the following notation:

$$\int_1^2 (x^2 + 2)\,dx = \left[\frac{x^3}{3} + 2x\right]_1^2$$

(Inside the square brackets is put the indefinite integral, $F(x)$, but without the arbitrary constant.)

We evaluate the function inside the square bracket at the upper limit, then at the lower limit and subtract the latter from the former.

$$\int_1^2 (x^2 + 2)\,dx = \left[\frac{x^3}{3} + 2x\right]_1^2$$
$$= (\tfrac{8}{3} + 4) - (\tfrac{1}{3} + 2)$$
$$= 6\tfrac{2}{3} - 2\tfrac{1}{3}$$
$$= 4\tfrac{1}{3}.$$

Remark

We omit the arbitrary constant from inclusion in the square bracket because it is superfluous – if included it would always cancel out.

Example

To find $\int_1^4 (x^2 - 2x + 1)\,dx$.

$$\int_1^4 (x^2 - 2x + 1)\,dx = \left[\frac{x^3}{3} - x^2 + x\right]_1^4$$
$$= (\tfrac{64}{3} - 16 + 4) - (\tfrac{1}{3} - 1 + 1)$$
$$= (21\tfrac{1}{3} - 16 + 4) - (\tfrac{1}{3} - 1 + 1)$$
$$= (9\tfrac{1}{3}) - (\tfrac{1}{3})$$
$$= 9.$$

7.5.1 Exercises

Evaluate the following:

1. $\int_0^1 (1 + 3x)\,dx$

2. $\int_0^1 (1 - 3x^2)\,dx$

3. $\int_{-1}^2 (x^2 + x - 2)\,dx.$

7.6 Integration and areas under curves

The definite integral defined in the previous section may appear to be rather contrived with no apparent practical use. In fact it is precisely because it does have a practical use that it was introduced. The raison d'être for the definite integral is provided by the following proposition, sometimes referred to as the *fundamental theorem of calculus*.

Given a function $f(x)$ and two values of x, $x = a$ and $x = b$, then the *area* between the graph of $f(x)$, the x axis and the values $x = a$ and $x = b$, is given by $\int_a^b f(x)dx$, where this definite integral is as defined in section 7.5.

The area concerned is illustrated on the graph (Fig. 7.2) as the shaded part.

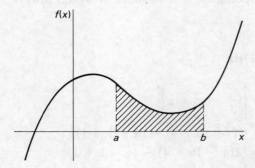

Fig. 7.2

Example
To find the area between the graph of $f(x) = x^2 + x + 2$, the x axis and $x = 1$ and $x = 3$, we evaluate $\int_1^3 (x^2 + x + 2)dx$ as outlined in section 7.5.

$$\int_1^3 (x^2 + x + 2)dx = \left[\frac{x^3}{3} + \frac{x^2}{2} + 2x\right]_1^3$$

$$= \left(\frac{27}{3} + \frac{9}{2} + 6\right) - \left(\frac{1}{3} + \frac{1}{2} + 2\right)$$

$$= 19\tfrac{1}{2} - 2\tfrac{5}{6}$$

$$= 16\tfrac{2}{3}.$$

The area is shown in the following diagram and according to our definite integral this area is $16\frac{2}{3}$ square units (Fig. 7.3).

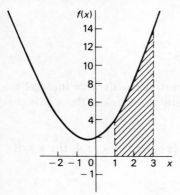

Fig. 7.3

Example

Find the area between the graph of $f(x) = 1 - x$, the x axis and $x = -1$, $x = 1$.

The area concerned is illustrated in the diagram (Fig. 7.4),

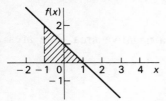

Fig. 7.4

and is given by $\displaystyle\int_{-1}^{1} (1-x)\,dx = \left[x - \frac{x^2}{2}\right]_{-1}^{1}$

$\qquad\qquad\qquad\quad = (1 - \tfrac{1}{2}) - (-1 - \tfrac{1}{2})$
$\qquad\qquad\qquad\quad = \tfrac{1}{2} - (-1\tfrac{1}{2})$
$\qquad\qquad\qquad\quad = \tfrac{1}{2} + 1\tfrac{1}{2}$
$\qquad\qquad\qquad\quad = 2$ square units.

7.6.1 Exercises

Find the areas between the graphs of the following functions, the x axis and the x values given.

1. $f(x) = 2 + x$ and between $x = 0$ and $x = 2$

2. $f(x) = x^2 - x + 1$ and between $x = 1$ and $x = 3$
3. $f(x) = 4 - x - x^2$ and between $x = -1$ and $x = 1$.

7.7 Areas below the x axis

Certain difficulties can arise in interpreting the definite integral as an *area*, if the area concerned happens to be below the x axis, as the following example illustrates.

Example

Find the area between the graph of $f(x) = x^2 - x - 4$, the x axis, and between $x = 0$, and $x = 2$.

The area will be given by

$$\int_0^2 (x^2 - x - 4)\,dx = \left[\frac{x^3}{3} - \frac{x^2}{2} - 4x\right]_0^2$$

$$= (\tfrac{8}{3} - \tfrac{4}{2} - 8) - (0)$$

$$= 2\tfrac{2}{3} - 2 - 8$$

$$= -7\tfrac{1}{3}.$$

This appears rather odd – having a negative area. The area found is shown in Fig. 7.5.

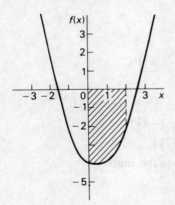

Fig. 7.5

As can be seen the area is *below* the x axis, and it is generally true that whenever the definite integral is used to calculate an area below the x axis, the definite integral produces a negative number.

In the above example this causes no problem, we can just ignore the negative sign and conclude that the area is $+7\frac{1}{3}$ square units. (Although in many economic examples we do in fact wish the answer to stay negative. If, for example, $f(x)$ were an investment function (see section 7.8.5) a negative accumulation of capital does have a meaning.)

If the area under consideration lies partly above and partly below the x axis, the use of definite integrals will mean that there will be some 'cancelling out' between the area above the axis (which turns out positive) and that below (which turns out negative). This may be all right in some examples (like, for instance, the investment function example mentioned), but it may lead to an incorrect answer in other cases. If the area is required *without* any cancelling (sometimes referred to as the *absolute area*) then the positive and negative areas must be calculated separately.

Example

Find the area between $f(x) = x^2 + x - 2$, the x axis and $x = 0$, $x = 2$.

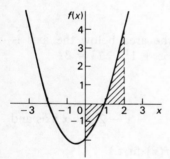

Fig. 7.6

Integration between $x = 0$ and $x = 2$ yields

$$\int_0^2 (x^2 + x - 2)\,dx = \left[\frac{x^3}{3} + \frac{x^2}{2} - 2x\right]_0^2$$

$$= \left(\frac{8}{3} + \frac{4}{2} - 4\right) - (0)$$

$$= 2\tfrac{2}{3} + 2 - 4$$

$$= \tfrac{2}{3}.$$

There has been some 'cancelling out' of the area below the x axis with that above the x axis (Fig. 7.6).

If this cancelling is *not* desired then the two areas must be calculated separately.

The area *below* the x axis $= \int_0^1 (x^2 + x - 2)\,dx$

$$= \left[\frac{x^3}{3} + \frac{x^2}{2} - 2x\right]_0^1 = -1\tfrac{1}{6}.$$

The area *above* the x axis $= \int_1^2 (x^2 + x - 2)\,dx$

$$= \left[\frac{x^3}{3} + \frac{x^2}{2} - 2x\right]_1^2 = \left(\frac{8}{3} + \frac{4}{2} - 4\right)$$
$$- \left(\frac{1}{3} + \frac{1}{2} - 2\right)$$
$$= 1\tfrac{5}{6}.$$

If we are to ignore the fact that the area below the axis is negative then the absolute area will be $+ 1\tfrac{1}{6} + 1\tfrac{5}{6} = 3$.

7.7.1 Exercise

Find the *absolute* area between $f(x) = x^2 - x - 2$, the x axis and $x = 1, x = 3$.
(Hint: You must draw the graph of $f(x)$ first.)

7.8 Applications to economics

7.8.1 Marginal cost and total cost

In section 5.6.1 we defined the marginal cost function as being the derivative of the total cost function.
Example
If cost $= x^3 - 12x^2 + 60x + 20$,
then marginal cost $= 3x^2 - 24x + 60$.

We are now in a position to perform the reverse operation, i.e. given a marginal cost function we can find the total cost function (and hence the average cost function) by integration.

Example

If the marginal cost function for a firm is given by
   marginal cost = $3x^2 - 20x + 40$
then the total cost function is obtained by integrating the marginal cost function,

i.e. total cost = $\int (3x^2 - 20x + 40)\, dx$

$\qquad\qquad = x^3 - 10x^2 + 40x + c$

(using the rules in sections 7.2.1 and 7.2.3).

N. B. The arbitrary constant of integration is equal to the fixed cost. It is impossible to determine this value from the marginal cost function only.

7.8.2 Exercise

A firm is faced with the following marginal cost function:
   marginal cost = $3x^2 - 10x + 30$.
If the firm's fixed cost is 50, find the total and average cost functions.

7.8.3 Marginal revenue and total revenue

The marginal revenue function was defined in section 5.6.3 as the derivative of the total revenue function.

Example

A firm has a total revenue function given by:
   revenue = $100x - 4x^2$.
Then   marginal revenue = $100 - 8x$.

As in section 7.8.1 we can again reverse the operation, i.e. given a marginal revenue function we can find the total revenue function (and hence the demand function) by integration.

Example

The marginal revenue function for a firm is given by
   marginal revenue = $250 - 10x$.

The total revenue function is then given by

$$\text{total revenue} = \int (250 - 10x)\,dx$$

$$= 250x - 5x^2 + c \text{ (using the rules in sections 7.2.1 and 7.2.3)}.$$

In the case of the total revenue function we can usually argue that if the firm produces no output then it will receive no revenue, i.e. we have one ordered pair for the total revenue function and this is sufficient to determine the arbitrary constant (see section 7.3).

When $x = 0$ the total revenue $= 250 \cdot 0 - 5 \cdot 0 + c$
$= c$.

But when $x = 0$ the total revenue $= 0$.

Therefore $c = 0$ and the total revenue function is
  total revenue $= 250x - 5x^2$.

From this function we can also find the demand function the firm is faced with. Since revenue $= p \cdot x$ we have
$p \cdot x = 250x - 5x^2$
and dividing both sides by x gives the demand function
$p = 250 - 5x$.

7.8.4 Exercise

If a firm's marginal revenue function is
  marginal revenue $= 20 - x$
find the total revenue function and the demand function.

7.8.5 Investment and capital stock

Consider a firm which is investing continuously in new machines, and suppose that at any point in time t, the rate of net investment I, is given by the function.
  $I = -t^2 + 4t + 8$.
In other words if we choose any point in time, say $t = 4$, the rate of net investment at that time can be obtained from the function, e.g. when $t = 4$, $I = -4^2 + 4 \cdot 4 + 8 = 8$.

We are assuming that investment is carried out *continuously*, i.e. at *any* point in time investment is being made.

For example, at $t = 4 \quad I = 8$
$\phantom{\text{For example, at }} t = 4.5 \quad I = 5.75$
$\phantom{\text{For example, at }} t = 4.75 \quad I = 4.4375$.

We can represent this 'investment function' on a graph (Fig. 7.7).

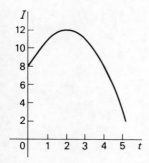

Fig. 7.7

The area under the curve between any two values of t represents the capital accumulated during that period.

Suppose we now wish to find how much capital has been accumulated from, say, $t = a$ to $t = b$. The capital accumulated in this time is given by

$$\int_a^b I \, dt.$$

In the previous example, the firm would accumulate capital equal to

$$\int_0^3 (-t^2 + 4t + 8) \, dt$$

in the first three years, and

$$\int_0^3 (-t^2 + 4t + 8) \, dt = \left[-\frac{t^3}{3} + 2t^2 + 8t \right]_0^3$$
$$= (-9 + 18 + 24) - (0) = 33 \text{ units}.$$

So the firm accumulates 33 units of capital in the first three years.

7.8.6 Exercise

A firm undertakes net investment according to the following investment function:
$$I = 10 + 20t - t^2.$$
Draw the graph of the function.
Find the capital accumulated during the time $t = 0$ to $t = 3$ and that accumulated from $t = 3$ to $t = 6$.

Answers to exercises

1.4.1   1. (a) $A \cup B = \{1, 2, 3, 5\}$
(b) $A \cap B = \{2, 3\}$
(c) $A \cap C = \emptyset$
(d) $A \cup B \cup C = \{1, 2, 3, 4, 5, 7, 9\}$
(e) $A \cup (B \cap C) = \{1, 2, 3, 5\}$
(f) $A \cap (B \cup C) = \{2, 3\}$

2. (a) True
(b) False
(c) True.

3. (a) False
(b) True
(c) True
(d) False.

4. $A \cup B = A$
$A \cap B = B$.

5. (a) $W \cup Z$ consists of people either unemployed or over 60 years of age.
(b) $W \cap X$ is the set of unemployed males.
(c) $Y \cap W \cap X$ is the set of unemployed males over 21.
(d) John Brown is unemployed.
(e) There is at least one male over 60.
(f) Everyone over 60 years of age is also over 21 years of age.
(g) Jane Brown is not unemployed.
(h) $(X \cap W) \cup (X \cap Z)$ consists of males who are unemployed together with males over 60.
(i) $X \cap (W \cup Z)$ is the set of people who are male and either unemployed or over 60 (which is in fact the same as the previous set).

1.5.1

1. (a)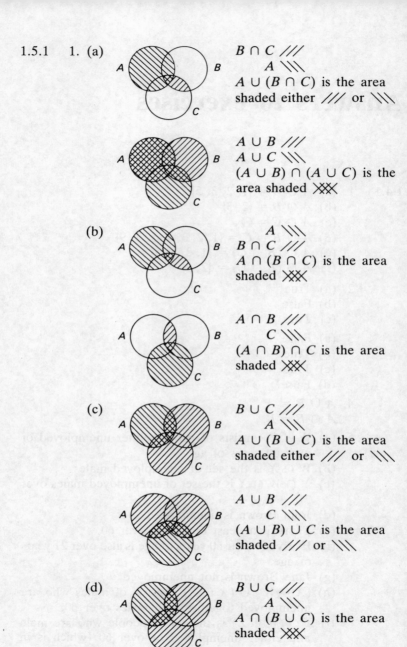

$B \cap C$ ///
A \\\
$A \cup (B \cap C)$ is the area shaded either /// or \\\

$A \cup B$ ///
$A \cup C$ \\\
$(A \cup B) \cap (A \cup C)$ is the area shaded ✕✕✕

(b)

A \\\
$B \cap C$ ///
$A \cap (B \cap C)$ is the area shaded ✕✕✕

$A \cap B$ ///
C \\\
$(A \cap B) \cap C$ is the area shaded ✕✕✕

(c)

$B \cup C$ ///
A \\\
$A \cup (B \cup C)$ is the area shaded either /// or \\\

$A \cup B$ ///
C \\\
$(A \cup B) \cup C$ is the area shaded /// or \\\

(d)

$B \cup C$ ///
A \\\
$A \cap (B \cup C)$ is the area shaded ✕✕✕

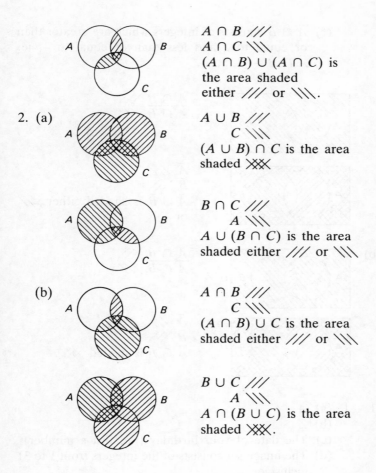

$A \cap B$ ///
$A \cap C$ \\\
$(A \cap B) \cup (A \cap C)$ is the area shaded either /// or \\\.

2. (a)

$A \cup B$ ///
C \\\
$(A \cup B) \cap C$ is the area shaded ✕✕✕

$B \cap C$ ///
A \\\
$A \cup (B \cap C)$ is the area shaded either /// or \\\

(b)

$A \cap B$ ///
C \\\
$(A \cap B) \cup C$ is the area shaded either /// or \\\

$B \cup C$ ///
A \\\
$A \cap (B \cup C)$ is the area shaded ✕✕✕.

1.6.1 1. (a) $\bar{A}$ consists of integers greater than or equal to 2
   (b) $\bar{B}$ consists of integers less than or equal to 4
   (c) $\overline{A \cup B}$ consists of integers which are not less than 2 or greater than 4, i.e. $\overline{A \cup B} = \{2, 3, 4\}$
   (d) $\bar{A} \cup \bar{B}$ consists of integers which are either greater than or equal to 2 or less than or equal to 4 (in fact $\bar{A} \cup \bar{B}$ consists of all integers of $\bar{A} \cup \bar{B} = U$)
   (e) $\overline{A \cap B}$ consists of integers which are not less than 2 and greater than 4 (in fact none of the integers are less than 2 and greater than 4 so $\overline{A \cap B} = U$).

(f) $\bar{A} \cap \bar{B}$ consists of integers which are greater than or equal to 2 and less than or equal to 4, ie. $\bar{A} \cap \bar{B} = \{2, 3, 4\}$.

2. (a)

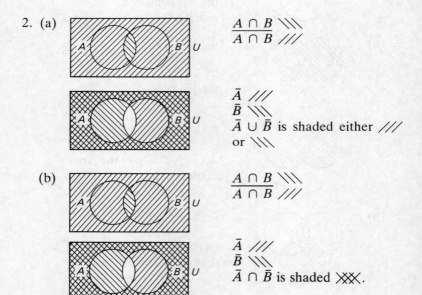

(b)

2.2.1 1. (a) A
   (b) Z
   (c) The date of your birthday (i.e. just a number)
   (d) The image set consists of the integers from 1 to 31 inclusive.

2. (a) The image of 2 is 3, the image of 0 is -1, the image of -5 is 24
   (b) The image set is $\{x|x$ is a real number and $x \geq -1\}$.

2.4.1 1. (a) The image of 0 is 2, the image of 2 is 10, the image of -1 is -2, the image of 4 is 18
   (b) The mapping is 'one to one' and therefore a function.

2. (a) The image of 1 is 1, the image of 3 is 17, the image of -1 is 1.

(b) The mapping is 'many to one' and therefore a function.

2.5.1

1.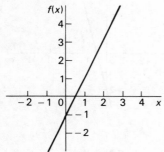

2.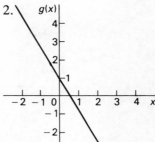

3.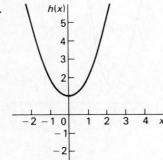

4.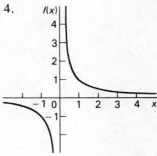

5.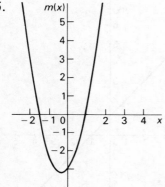

2.7.5   1. Linear

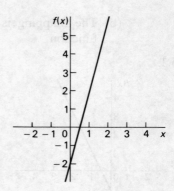

2. Quadratic

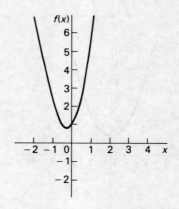

3. Linear

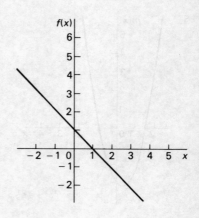

4. Quadratic

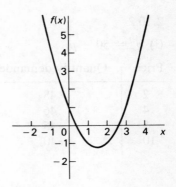

5. None

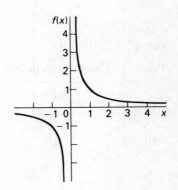

6. Third degree polynomial

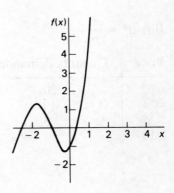

2.10.2

(i) $q^d = 50 - p$

Price	Quantity demanded
2	48
4	46
5	45
10	40

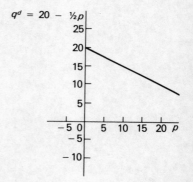

(ii) $q^d = 20 - \frac{1}{2}p$

Price	Quantity demanded
2	19
4	18
5	$17\frac{1}{2}$
10	15

(iii) $q^d = \dfrac{40}{p}$

Price	Quantity demanded
2	20
4	10
5	8
10	4

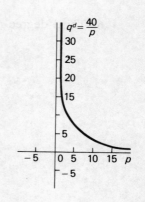

2.10.5 (i) $q^s = 1 + 3p$

Price	Quantity supplied
2	7
4	13
5	16
10	31

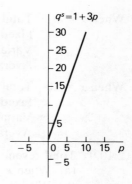

(ii) $q^s = 2 + \frac{1}{2}p$

Price	Quantity supplied
2	3
4	4
5	$4\frac{1}{2}$
10	7

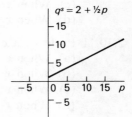

(iii) $q^s = \log_e p$

Price	Quantity supplied
2	0.693 1
4	1.386 3
5	1.609 4
10	2.302 6

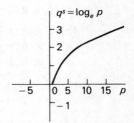

2.10.7 Total cost $= x^3 - 10x^2 + 150x + 200$
Fixed cost $= 200$
Variable cost $= x^3 - 10x^2 + 150x$
Average cost $= x^2 - 10x + 150 + \dfrac{200}{x}$

When $x = 1$ Total cost $= 341$
Fixed cost $= 200$
Variable cost $= 141$
Average cost $= 341$.

When $x = 5$ Total cost $= 825$
Fixed cost $= 200$
Variable cost $= 625$
Average cost $= 165$.

When $x = 10$ Total cost $= 1\,700$
Fixed cost $= 200$
Variable cost $= 1\,500$
Average cost $= 170$.

2.10.9 Total revenue $= 40x - 2x^2$
(a) When $x = 0$, revenue $= 0$
(b) When $x = 2$, revenue $= 72$
(c) When $x = 10$, revenue $= 200$
(d) When $x = 20$, revenue $= 0$.

2.10.11 Profit $= 60x + 9x^2 - x^3 - 200$
(a) When $x = 0$, profit $= -200$
(b) When $x = 2$, profit $= -52$
(c) When $x = 10$, profit $= 300$
(d) When $x = 20$, profit $= -3\,400$.

3.2.5 (a) $f + g : x \to x^2 + 3x + 2$
$f + g : 1 \to 6$

(b) $f \times h : x \to (3x + 2)(1 - x)$
or $f \times h : x \to 2 + x - 3x^2$
$f \times h : 1 \to 0$

(c) $g - h : x \to x^2 - 1 + x$
$g - h : 1 \to 1$

(d) $f + g + h : x \to x^2 + 2x + 3$
$f + g + h : 1 \to 6$

(e) $f \times (g + h) : x \to (3x + 2)(x^2 + 1 - x)$
or $f \times (g + h) : x \to 3x^3 - x^2 + x + 2$
$f \times (g + h) : 1 \to 5$

(f) $(f \times g) + h : x \to 3x^3 + 2x^2 + 1 - x$
$(f \times g) + h : 1 \to 5$

(g) $(f - h) \div g : x \to \dfrac{4x + 1}{x^2}$

$(f - h) \div g : 1 \to 5$.

3.3.1 $f + g : 0 \to 2$
$f \times g : 0 \to 0$
$g - f : 0 \to -2$
$gof : 0 \to 6$
$fog : 0 \to 2$

$f + g : 1 \to 6$
$f \times g : 1 \to 9$
$g - f : 1 \to 0$
$gof : 1 \to 9$
$fog : 1 \to 5$
$f + g : 2 \to 10$
$f \times g : 2 \to 24$
$g - f : 2 \to 2$
$gof : 2 \to 12$
$fog : 2 \to 8$.

3.4.1   (a) $gof : x \to 5 + 12x$
(b) $gof : x \to 64x^2$
(c) $gof : x \to 10x^2 + 3$.

3.5.1   1. (a) $fog : x \to 12x + 9$
(b) $fog : x \to 8x^2$
(c) $fog : x \to 5(2x + 3)^2$
or $fog : x \to 20x^2 + 60x + 45$

2. (a) $f(x) = g(x) + h(x)$
(b) $f(x) = \dfrac{h(x)}{g(x)}$
(c) $f(x) = h(x) \cdot g(x)$
(d) $f(x) = h(g(x))$
(e) $f(x) = g(h(x))$.

4.2.3   1. (a) $x = -2$
(b) $x = 4$
(c) $x = -\frac{1}{6}$
(d) $x = 4$
(e) $x = 9$.

4.3.4   1. $x = -2, \quad x = -1$
2. $x = 3, \quad x = -1$
3. $x = 3, \quad x = -\frac{1}{2}$
4. $x = 1, \quad x = -1$.

4.4.4   1. Two real solutions
        2. One real solution
        3. No real solutions.

4.6.3 $x = \frac{1}{2}$, $x = \frac{3}{2}$, $x = -\frac{1}{2}$.

4.7.2   1. $x \leq \frac{1}{2}$
        2. $x \geq 4$
        3. $x \geq -\frac{3}{2}$
        4. $x \leq -\frac{3}{2}$

4.7.5   1. $-2 \leq x \leq 1$
        2. $x \leq -2$ or $x \geq 1$
        3. $x \leq 1$ or $2 \leq x \leq 3$
        4. No solution.

4.8.4   1. $x = 2$
        2. $x = 2$
        3. $x = 1$, $x = -2$
        4. No solution.

4.8.6   1. $x = 2$, $y = 8$
        2. $x = 2$, $y = 7$
        or $x = -3, y = 2$.

4.8.8   1. $x = 0$, $y = 6$, $z = 13$
        2. $x = 0$, $y = 1$, $z = 2$.

4.9.5   1. The new equilibrium price is $p = \dfrac{a-c}{d-b} - \dfrac{ds}{d-b}$

        2. The new equilibrium price is $p = \dfrac{a-c}{d-b} + \dfrac{x}{d-b}$.

5.5.4   1. $f'(x) = 2x - 3$
           When $x = 0$ the slope of $f(x)$ is -3
           When $x = 1$ the slope of $f(x)$ is -1
           When $x = -1$ the slope of $f(x)$ is -5

        2. $f'(x) = -1 - 8x + 3x^2$
           When $x = 0$ the slope of $f(x)$ is -1
           When $x = 1$ the slope of $f(x)$ is -6
           When $x = -1$ the slope of $f(x)$ is 10

3. $f'(x) = 3$
   When $x = 0$ the slope of $f(x)$ is 3
   When $x = 1$ the slope of $f(x)$ is 3
   When $x = -1$ the slope of $f(x)$ is 3.

5.5.7 1. $f'(x) = 2x(3x - 2) + 3(1 + x^2) = 9x^2 - 4x + 3$
2. $f'(x) = 3x^2(1 - x) + (-1)(x^3) = 3x^2 - 4x^3$
3. $f'(x) = \dfrac{(2x)(1 - x) - (-1)(x^2)}{(1 - x)^2} = \dfrac{2x - x^2}{(1 - x)^2}$
4. $f'(x) = \dfrac{2x(1 - 3x) - (-3)(x^2 + 2)}{(1 - 3x)^2} = \dfrac{2x - 3x^2 + 6}{(1 - 3x)^2}.$

5.5.10 1. (a) $f'(x) = e^x$
   When $x = 0$ the slope of $f(x)$ is 1
   When $x = 1$ the slope of $f(x)$ is 2.718

   (b) $f'(x) = \cos(x)$
   When $x = 0$ the slope of $f(x)$ is 1
   When $x = 1$ the slope of $f(x)$ is 0.54

   (c) $f'(x) = -\sin(x)$
   When $x = 0$ the slope of $f(x)$ is 0
   When $x = 1$ the slope of $f(x)$ is -0.841

2. $f'(x) = \dfrac{1}{x}$

   When $x = 1$ the slope of $f(x) = 1$
   When $x = 0$ $f'(x)$ is not defined (we cannot have $\frac{1}{0}$)
   and so the slope of $f(x)$ cannot be found when $x = 0$

3. (a) $f'(x) = 2xe^{x^2+1}$
   (b) $f'(x) = -3e^{-3x}$
   (c) $f'(x) = \dfrac{2}{2x - 1}$

   (d) $f'(x) = -4\cos(1 - 4x)$
   (e) $f'(x) = -2x\sin(x^2)$
   (f) $f'(x) = 2\sin x \cos x - 2\cos x \sin x = 0.$

5.6.2 Cost $= x^3 - 4x^2 + 10x + 200$
Marginal cost $= 3x^2 - 8x + 10$

Average cost $= x^2 - 4x + 10 + \dfrac{200}{x}$

(a) When $x = 1$ Total cost $= 207$
Marginal cost $= 5$
Average cost $= 207$

(b) When $x = 2$ Total cost $= 212$
Marginal cost $= 6$
Average cost $= 106$

(c) When $x = 10$ Total cost $= 900$
Marginal cost $= 230$
Average cost $= 90$.

5.6.4 Total revenue $= 50x - 5x^2$
Marginal revenue $= 50 - 10x$

(a) When $x = 0$ Price $= 50$
Total revenue $= 0$
Marginal revenue $= 50$

(b) When $x = 2$ Price $= 40$
Total revenue $= 80$
Marginal revenue $= 30$

(c) When $x = 5$ Price $= 25$
Total revenue $= 125$
Marginal revenue $= 0$

(d) When $x = 10$ Price $= 0$
Total revenue $= 0$
Marginal revenue $= -50$.

5.6.6 The equilibrium price is $p = (a - c)/(d - b) - ds/(d - b) - f(s)$ and the corresponding quantity demanded and supplied

$$q^d = q^s = \frac{da - bc}{d - b} - \frac{bds}{d - b} = g(s).$$

The effect on the equilibrium price of a change in s is given by $f'(s) = -d/(d - b) < 0$,
i.e. an increase in the subsidy will result in a decrease in price.

The effect on the quantity demanded and supplied, of a change in s is given by $g'(s) = -bd/(d - b) > 0$,
i.e. an increase in the subsidy will result in an increase in the quantity sold.

6.2.2 1. $x = -4$
2. $x = \frac{3}{2}$.

6.2.4 1. $x = -4$
2. $x = \frac{3}{2}$.

6.3.1 1. $x = -2$ is a minimum
2. $x = -\frac{1}{2}$ is a maximum
3. $x = -1$ is a maximum
$x = +1$ is a minimum
4. $x = -2$ is a point of inflexion.

6.4.1 1. $f'(x) = 4x^3 - 6x$
$f''(x) = 12x^2 - 6$

2. $f'(x) = -1$
$f''(x) = 0$

3. $f'(x) = 2x \cdot e^{x^2}$ (using section 5.5.8.)
$f''(x) = 2e^{x^2} + 2x \cdot 2xe^{x^2}$ (using section 5.5.5)
$= 2e^{x^2}(1 + 2x^2)$

4. $f'(x) = -\sin(x)$
$f''(x) = -\cos(x)$.

6.5.3 1. $x = 3$ is a minimum
2. $x = \frac{1}{2}$ is a maximum
3. $x = -3$ is a maximum
$x = -1$ is a minimum
4. $x = 0$ is a maximum
5. $x = -1$ is a point of inflexion.

6.6.2 Total revenue $= 100x - \frac{1}{2}x^2$
profit $= 90x - x^2 - 50$
Profit is maximised when $x = 45$. The profit is then $1\,975$.

6.6.4 Marginal cost = $x + 10$
Marginal revenue = $100 - x$
When $x = 45$ marginal cost = 55 and marginal revenue = 55.

6.6.6 Average cost = $2x^2 - 20x + 100$
Marginal cost = $6x^2 - 40x + 100$
Average cost is minimised when $x = 5$
When $x = 5$ average cost = 50 and marginal cost = 50.

7.3.1
1. $x - \dfrac{x^2}{2} + c$

2. $\dfrac{x^3}{3} - \dfrac{3x^2}{2} + 2x + c$

3. $-\tfrac{1}{2}x^{-2} + c$

4. $4x + c$

5. $\dfrac{x^4}{4} - x^3 + 2x^2 - 7x - \dfrac{1}{x} + c.$

7.4.3
1. $-x \cos x + \sin x + c$
2. $xe^x - e^x + c$
3. $x^2 e^x - 2xe^x + 2e^x + c$
4. $\tfrac{1}{3}(x^3 + 2)^3 + c$
5. $\log_e (x^2 + x + 1) + c$
6. $e^{x^2} + c.$

7.5.1
1. $\tfrac{5}{2}$
2. 0
3. $-1\tfrac{1}{2}.$

7.6.1
1. 6 square units
2. $6\tfrac{2}{3}$ square units
3. $7\tfrac{1}{3}$ square units.

7.7.1 3 square units.

7.8.2 Total cost = $x^3 - 5x^2 + 30x + 50$

average cost = $x^2 - 5x + 30 + \dfrac{50}{x}.$

7.8.4 Total revenue = $20x - \dfrac{x^2}{2}$

Demand function is $p = 20 - \dfrac{x}{2}$.

7.8.6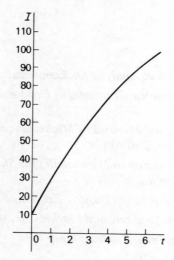

Capital accumulated from $t = 0$ to $t = 3$ is 111 units.
Capital accumulated from $t = 3$ to $t = 6$ is 237 units.

Bibliography

1. Allen, R. G. D. — *Mathematical Analysis for Economists.*
2. Archibald, G. C. & Lipsey, R. G. — *A Mathematical Treatment of Economics.*
3. Chiang, A. C. — *Fundamental Methods of Mathematical Economics,* 2nd edn.
4. Lewis, J. Parry — *An Introduction to Mathematics for Students of Economics.*
5. Lipsey, R. G. — *An Introduction to Positive Economics.*
6. Pickering, J. F. — *Industrial Structure and Market Conduct.*
7. Walsh, V. C. — *Introduction to Contemporary Microeconomics.*
8. Yamane, T. — *Mathematics for Economists.*

Index

absolute area, 183
addition of functions, 66–7
arbitrary constant of integration, 167–70
areas under curves, 180–4
  absolute, 183
  below x axis, 182–4
average cost, 60
  relation to marginal cost, 159–61
average revenue, 62

BEDMAS, 9

calculus, fundamental theorem of, 180
capital stock and investment, 186–8
codomain of set, 21
complement of set, 16–18
complex numbers, 90–2
composition of functions, 68–73
  derivation of formula for fog, 72–3
  derivation of formula for gof, 70–2
conjugate pairs, 91
constant functions, 37–8
constant of integration, 167–70
continuous functions, 122
cosine function, 51–2
  differentiation, 130
  integration, 167
cost
  average *see* average cost
  fixed, 60
  marginal *see* marginal cost
  variable, 60
cost functions, 58–61
  total, 59

definite integrals, 177–9
  notation, 178–9
demand functions, 53–6
  notation, 54

  price form of, 56
  quantity form of, 56
derivatives, 121
  second and higher, 148–9
differentiation, 121–37
  definition, 121–2
  function of a function rule, 128–9
  of differences, 125–6
  of products, 126–7
  of quotients, 127–8
  of sums, 125
  power rule, 123–4
  rules of, 123–32
discrete functions, 123
disjoint sets, 7
division of functions, 66–7
domain of set, 21

e (base of natural logarithms), 46
element of set, 2
empty set, 5
equations, 75–114
  higher degree polynomial, 92–3
  linear *see* linear equations
  quadratic *see* quadratic equations
  simultaneous *see* simultaneous equations
equilibrium in simple market model, 107–8
exponential function(s), 45–7
  differentiation, 130
  integration, 167
  natural, 47
exponential growth, 47
exponents, 43–5

factorisation, 82
fixed cost, 60
fractions, 3
function(s)
  addition, 66–7

function(s) (*continued*)
  composition *see* composition of
    functions
  constant, 37–8
  continuous, 122
  cosine, 51–2
  cost *see* cost functions
  definition, 25
  demand *see* demand functions
  discrete, 123
  division of, 66–7
  exponential *see* exponential
    function(s)
  graph of *see* graphs
  inverse, 42
  linear, 36–7
  logarithm *see* logarithm functions
  maximisation of *see* maximisation
    of functions
  minimisation of *see* minimisation
    of functions
  multiplication of, 66
  polynomial, 40
  profit *see* profit functions
  quadratic, 38–40
  revenue, 61–2
  sine, 50–1
  slope of *see* slope of function
  subtraction of, 66
  supply *see* supply functions
function of a function (*see also*
    composition of functions)
  in differentiation, 128–9

global maximum point, 139
global minimum point, 139
graphs, 28–35
  slope *see* slope of function

image set, 22–3
imaginary numbers, 4, 91
  conjugate pairs, 91
indefinite integrals (*see also*
    integration), 177
inequalities, 93–7
  linear, 94–6
    solution by graph, 94–5
    solution by manipulation, 95–6
  non-linear, 96–7
  strict, 96
inflexion, point of, 147

integers, 3
  negative, 3
  positive, 2
integrals, definite *see* definite integrals
integration, 162–88
  arbitrary constant of, 167–70
  areas under curves and, 180–4
  as reverse of differentiation, 162
  by parts, 170–4
  limits of, 178
  notation, 162–3
  of differences, 166
  of sums, 165–6
  power rule, 164–5
  rules of, 163–7
  substitution rule, 174–6
intercept of graph, 36
intersection of sets
  definition, 6–7
  Venn diagram representation, 12
inverse function, 42
investment and capital stock, 186–8
irrational numbers, 3

limits of integration, 178
linear equations, 76–80
  solution by graph, 76–8
  solution by manipulation, 78–9
linear functions, 36–7
  intercept on axis, 36
  slope of graph, 36
local maximum point, 139
local minimum point, 139
logarithm functions
  differentiation, 130–1
  natural, 48–9
  properties of, 49–50

many-to-many mapping, 25
many-to-one mapping, 24
mapping(s)
  definition, 21
  notation, 21–2, 26–8
  operations on, 65–74
  reverse, 41–2
  types of, 23–5
marginal cost, 132–4
  relation to average cost, 159–61
  relation to marginal revenue, 157–9
  relation to total cost, 184–5
marginal revenue, 134–6
  relation to marginal cost, 157–9
  relation to total revenue, 185–6

market models
  general linear, 110–11
    imposition of tax on, 111–13
  simple
    equilibrium in, 107–8
    imposition of tax on, 108–10, 136–7
maximum point, 138
  first order condition for, 140–3
  second order condition for, 149–50
minimum point, 139
  first order condition for, 143–5
  second order condition for, 150–2
models, market *see* market models
multiplication of functions, 66

natural exponential function, 47
natural logarithm function, 48–9
natural numbers, 2
negative integers, 3

one-to-many mapping, 24–5
one-to-one mapping, 25
operations on mappings, 65–74
ordered pairs, 28–9
  definition, 28

point of inflexion, 147
polynomial equations, higher degree, 92–3
  solution by factorisation, 92–3
  solution by graph, 92
polynomial functions, 40
positive integers, 2
power rule of differentiation, 123–4
power rule of integration, 164–5
profit functions, 63–4
profit maximisation, 154–7

quadratic equations, 80–92
  no real solutions, 89–90
  one real solution, 88–9
  solution by factorisation, 81–3
  solution by formula, 83–6
  solution by graph, 80–1
  two unequal real solutions, 86–8
  types of solution, 86–90
quadratic functions, 38–40

rate of change, 122
rational numbers, 3
real line, 29
real numbers, 2–4

revenue
  average, 62
  marginal *see* marginal revenue
revenue functions, 61–2
reverse mappings, 41–2

second derivatives, 148–9
set(s), 1–19
  codomain of, 21
  complement of, 16–18
  definition, 1–2
  disjoint, 7
  domain of, 21
  element of, 2
  empty, 5
  equality of, 4
  image, 22–3
  intersection *see* intersection of sets
  mapping, *see* mapping(s)
  operations on, 6–11
    order of, 9–10
  union *see* union of sets
  universal, 16
simultaneous equations, 97–106
  algebraic method, 101–3
  graph method, 98–101
  more than two unknowns, 105–6
  notation, 103–4
sine function, 50–1
  differentiation, 130
  integration, 167
slope of function, 36
  at a point, 117–18
  linear functions, 115–16
  method for finding, 118–21
  non-linear functions, 116–17
stationary points, 146–8
  identification of types, 153–4
subset, 4–6
  definition, 4
substitution rule in integration, 174–6
subtraction of functions, 66
supply functions, 56–8
  notation, 57

total cost (*see also* cost)
  relation to marginal cost, 184–5
total cost function (*see also* cost functions), 59
total revenue (*see also* revenue)
  relation to marginal revenue, 185–6
total revenue function (*see also* revenue functions), 61

union of sets, 6
  Venn diagram representation, 12
universal set, 16
unreal numbers *see* imaginary numbers

variable cost, 60
  linear functions, 115–16
  method for finding, 118–21
  non-linear functions, 116–17